AGROFORESTRY HANDBOOK

Agroforestry for the UK

Edited by
Ben Raskin and Simone Osborn

1st Edition
(July 2019)

Disclaimer

The information provided within this handbook is for general informational purposes only. While the authors and publishers have made every effort to ensure that the information in this book was correct at the time of publication, there are no representations or warranties, express or implied, about the completeness, accuracy, reliability, suitability or availability with respect to the information, products, services, or related graphics contained in this handbook for any purpose. Any use of this information is at your own risk.

Preface

The principles behind agroforestry are not new, they have been practised for as long as humans have been farming. Realising the full potential of integrating commercially managed trees into farming systems is, however, still a long way off. This book introduces the theory of agroforestry. We look at practical management and design considerations. Where the information is available we have provided information on markets and pricing. There are countless opportunities to grow a range of tree crops for human and animal consumption, for animal welfare and soil health, for building, and even woody products to replace plastics. The potential benefits are clear both to individual farms and to the wider environment, for instance by sequestering carbon and reducing flood risk.

The agroforestry journey is just beginning. A few pioneering farmers are planting commercial systems, and some are already seeing concrete commercial and environmental benefits, but trees grow slowly and we will learn more as these systems mature.

The authors are leading researchers and practitioners with decades of experience in agroforestry from around the world, including research here in the UK. We hope this handbook will give farmers and advisors the confidence and inspiration to start planting. The Soil Association and the Woodland Trust are delighted to be able to support the Farm Woodland Forum www.agroforestry.ac.uk who are the leading UK agroforestry body. Join them for further information and support on all things agroforestry.

Helen Chesshire – The Woodland Trust,
Ben Raskin – Soil Association.

Contents

Chapter 1 What is Agroforestry

10	What is agroforestry?
11	What are the types of agroforestry?
12	Silvopastoral agroforestry
15	Silvoarable agroforestry
16	Agrosilvopastoral systems
17	Hedgerows, shelterbelts and riparian buffer strips
18	Forest farming, Homegardens
18	Why agroforestry?

Chapter 2 Agroforestry systems design

19	Introduction and aims
20	Introduction to agroforestry design
21	The key elements of agroforestry design
22	The Purpose of your agroforestry project or intervention
23	Advice
23	Measures of success
27	Agroecology
28	Starting points
30	Adaptive management
32	Layout
35	Practical considerations when designing specific agroforestry systems in the UK
36	Practical design example silvopastoral
38	Practical design example silvoarable
40	Practical design example walnut trial
42	Woodland/forest systems including forest gardens
43	Hedgerow/buffer strip systems
43	Landscape, estates and partnerships
43	Observation on wildlife

Chapter 3 **Silvopasture**

45	What is silvopasture?
46	How can silvopasture benefit my farming system?
51	Different types of silvopasture system
53	Designing for livestock benefits
56	CASE STUDY: Trees mean better business
57	Maximising the value of the trees

Chapter 4 **Silvoarable**

61	What is silvoarable?
64	Types of silvoarable systems
69	CASE STUDY: Whitehall Farm – Planting to improve economic returns
70	Maximising the arable benefits
74	Maximising the value of the trees

Chapter 5 **Hedges, windbreaks and riparian buffers**

80	Site selection, design and establishment
81	Hedges
83	Hedgerows – The management cycle and planning
84	Hedgerows for woodfuel
85	Introducing a coppice cycle to hedges
86	Managing hedgerows for other products
88	Windbreaks
89	Windbreaks – Management for protection and production
90	Riparian buffers
91	Riparian buffers – Management for protection and production
92	CASE STUDY: Trees enhance flock health and field drainage
93	Legal and other considerations

Chapter 6 The economic case for agroforestry

95	Introduction
97	Implications of agroforestry design on farm economics
98	Financial evaluation of agroforestry
99	Whole farm budgeting for profit from agroforestry
106	Agroforestry fixed costs, labour and machinery budgeting
108	Gross margin analysis of agroforestry
109	Silvoarable gross margins
112	Silvohorticulture gross margins
115	Lowland silvopasture gross margins
118	Forecasting agroforestry outputs and costs
118	Impact on outputs from tree and crop competition
122	Productive potential of agroforestry – Land Equivalent Ratio
125	Cash flow forecasting for agroforestry
126	Land tenure
130	Goverment support for agroforestry
131	Agroforestry for carbon capture
132	**Market opportunities –** Direct outputs from the tree component of agroforestry systems
133	Market considerations
134	Tree establishment, maintenance and management
137	Tree growth rate yield and products
139	Tree species selection
142	References
148	Authors
150	Acknowledgements

Chapter 1
What is Agroforestry?
Dr Paul Burgess, Cranfield University

Agriculture and forestry have often been treated as separate and distinct disciplines in colleges, universities and handbooks on farm management. However on the ground, most farmers manage land that combines agricultural production with trees that stand individually or in groups too small to be classified as woodland. In 2017, the Forestry Commission[1] estimated there were 742,000 hectares (ha) of trees that weren't in woodlands in Great Britain (in other words less than 0.1 ha). That is about 3.3% of the area of Great Britain, similar to the area of barley grown in the UK.

Intensive arable and livestock systems can produce high yields per unit area and labour, but they can also have negative environmental effects. The increased uptake of agroforestry in the UK can offer productivity, mitigation of climate change, water management, biodiversity and landscape, and welfare benefits.

Productivity: Providing shelter for livestock can increase daily live weight gain, and windbreaks in arable systems can reduce soil erosion. Trees can provide additional sources of on-farm revenue such as woodfuel, timber, fruit, or nuts.

Tacking climate change: Trees can moderate the local climate and the additional storage of carbon (above-ground and below-ground) can contribute to national greenhouse gas targets[2].

Water management: Including trees can help slow the flow of run-off from farms thereby moderating downstream flood flows and reducing soil erosion. The deep roots of appropriately placed trees can help minimise the leaching of nitrate.

Biodiversity and landscape enhancement: Across Europe, agroforestry on farmland has been shown to significantly increase the diversity of species, and there is increasing evidence that many people favour mosaic landscapes.

Animal welfare: Reduced temperature extremes and a greater variety of within-field habitats can reduce animal stress and allow more natural animal behaviour.

What is agroforestry?

A useful short definition of agroforestry is 'farming with trees'. Agroforestry includes both the integration of trees on farmland and the use of agricultural crops and livestock in woodlands. In the European Union, Article 23 of Regulation 1305/2013 defines agroforestry as "land use systems in which trees are grown in combination with agriculture on the same land". A more detailed definition is that agroforestry is "the practice of deliberately integrating woody vegetation (trees or shrubs) with crop and/or animal systems to benefit from the resulting ecological and economic interactions"[3]. Whereas 'mixed farming' is the integration of crops and livestock, agroforestry covers the area in a triangle where trees (and shrubs) are integrated with either crops, livestock, or mixed farming (Figure 1).

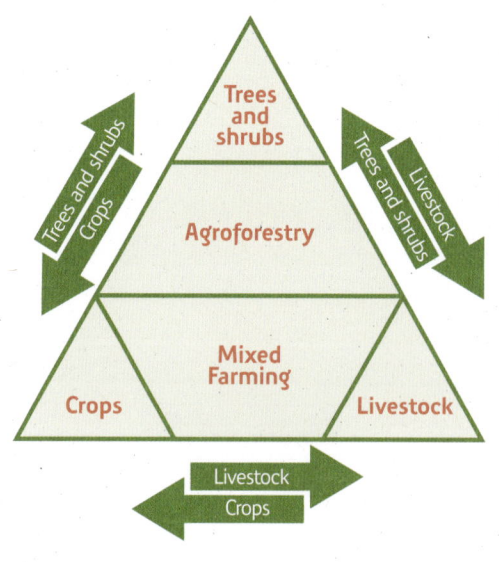

Figure 1: **Agroforestry involves the integration of trees and shrubs with either crop, livestock or mixed farming**

Patrick Worms at the UK 2017 Agroforestry Conference developed a quote by United States Supreme Court Justice Potter Stewart, in observing that agroforestry "is not easy to define but I know it when I see it". Whilst Patrick, who is the 2019 President of the European Agroforestry Federation (EURAF), may be able to recognise agroforestry, it is doubtful whether many others "know agroforestry when they see it".

This handbook argues for a wide definition of agroforestry. In fact, we argue that much of the British landscape is actually an agroforestry landscape comprising a mosaic of trees and farming systems.

Chapter 1
What is Agroforestry?

As part of a recent agroforestry research project called AGFORWARD, Rosa Mosquera-Losada et al. (2016)[4] identified five distinct types of agroforestry in Europe:

- Silvopastoral agroforestry: the combination of trees and livestock
- Silvoarable agroforestry: the combination of trees and crops
- Hedgerows, shelterbelts and riparian buffer strips
- Forest farming: crop cultivation within a forest environment, and
- Homegardens: combinations of trees and food production close to homes.

In this handbook we will focus on the first three.

A similar categorisation has also been proposed by Lawson et al. (2016)[5], including more detail on whether the agroforestry exists on agricultural or forest land, and whether it exists within or between fields (Table 1). The typology in Table 1 also distinguishes between trees within fields and trees between fields.

Table 1: **A typology for types of UK agroforestry developed from (Lawson et al. 2016)[4]**

	Agroforestry system	Official land use classification	
		Forest land	**Agricultural land**
Trees within fields	Silvopastoral	Forest grazing	Wood pasture Orchard grazing Individual trees
	Silvoarable	Forest farming	Alley cropping Alley coppice Orchard intercropping Individual trees
	Agrosilvopastoral	Mixtures of the above	
Trees between fields	Hedgerows, shelterbelts and riparian buffer strips	Forest strips	Shelterbelt networks Wooded hedges Riparian tree strips

The remainder of this chapter provides a brief introduction to the main types of agroforestry presented in Table 1 and examines the extent of such systems in the UK.

Silvopastoral agroforestry

Silvopastoral agroforestry is the integration of trees with livestock. As illustrated in Table 1 this can occur on forest land or agricultural land. It encompasses forest grazing, wood pasture, orchard grazing and newer forms of the integration of trees with livestock (Figure 2 and Figure 3).

Forest grazing

'Forest grazing' is where livestock is kept on land that is clearly designated as a forest and woodland. Rodwell and Patterson (1994)[6] report that "grazing and browsing by large herbivores are natural features of woodland ecosystems and grazing management should be considered from the outset, in management of semi-natural and native woods". In 2004, Lucy Sumsion and Meg Pollock developed a Woodland Grazing Kit[7] to guide woodland managers when it is appropriate to graze woodlands. Forest grazing tends to focus on the use of cattle, but it can also include sheep and pigs. For example, in some conservation systems, pigs are used to help establish new tree seedlings within an established woodland. The action of the pigs on the soil encourages dormant seeds to germinate. As woodland is the natural habitat of the ancestors of pigs, this is seen to be a historic system.

Figure 2: **Woodland grazing**

Pigs in a chestnut woodland in north-west Spain

©José Javier Santiago Freijanes

Wood pasture

Wood pasture is a wide encompassing term used to describe "landscapes in which livestock grazing co-occurs with scattered trees and shrubs"[8].

Figure 3: **Examples of silvopastoral systems include**

A Wood pasture
Extensive wood pasture systems with grazing cattle at Epping Forest in 2014

B Parkland system
Parkland system in Leicestershire in 2018

C Individually fenced alder trees with sheep grazing
Individually fenced nitrogen-fixing alder trees with sheep grazing at Henfaes in North Wales in 2012

D Individually fenced trees and hay production
Individually fenced trees combined with hay production near Louth Wood in Derbyshire in 2009

Photographs: ©Paul Burgess

In the UK, the term wood pasture is often combined with the term parkland. There is a UK Biodiversity Steering Group responsible for the conservation of wood pasture and parkland (https://ptes.org/wppn/) specifically focusing on sites with open-grown ancient or veteran trees with grazing livestock and an understorey of grassland or heathland[9].

From a conservation perspective, wood pastures are typically defined in terms of the presence of old trees. However, building on the work of the UK Silvopastoral National Network Experiment, new wood pasture systems have been developed. Examples of such systems include the production of high-value ash trees with sheep grazing at Loughgall in Northern Ireland, and the production of nitrogen-fixing alder with sheep grazing at Henfaes in North Wales.

Grazed orchards

The most established silvopastoral systems in the UK are grazed orchards (Figure 4). Analysis of Land use/cover area frame statistical survey (LUCAS) data suggests that there could be an area of 14,200 ha of grazed orchards in the UK[10].

Figure 4: **Grazed orchards**

A Apples with poultry
Apple trees with free-range hens in the Netherlands

B Apples with tree grazing
A high-stem grazed cider apple orchard in Herefordshire in 2017

Chapter 1
What is Agroforestry?

Silvoarable agroforestry

Silvoarable agroforestry integrates trees with arable crops, although the term is used more widely and includes short rotation coppice and horticultural crops (Figure 5). Compared to silvopastoral systems, the reported areas of silvoarable agroforestry are currently small (358,000 ha in Europe or 0.1% of territory) with the major areas in Europe occurring in Italy, Spain and Portugal.

Figure 5: Examples of silvoarable systems include

A Silvoarable alley cropping
Poplar with oilseed rape at Leeds University Farm, Yorkshire, 2003

B Alley coppice
An alley coppice system with wild cherry with willow short rotation coppice at Loughgall in Northern Ireland in 2016

C Hazel coppice system at Wakelyns
Planting of cereals between rows of hazel coppice at Wakelyns Agroforestry in Suffolk

D Alley cropping with vegetables
Use of tree rows to provide shelter for vegetable production at Shillingford Organics in Devon in 2014

Agrosilvopastoral systems

The term 'agrosilvopastoral' is used to describe agroforestry systems that combine trees, crops and livestock. In Spain, there are areas of oak rangeland where arable cropping and grazing is practised.

Temporal changes in agroforestry

At present, within the UK, we are not aware of any systems where crops, grazing and trees take place at exactly the same time. However there are some systems which are 'agrosilvopastoral' over the time period of a tree rotation. For example, Bill Acworth in Berkshire initiated a silvoarable system (Figure 6a), which he converted to a silvopastoral system as the tree canopies expanded (Figure 6b).

Figure 6: **A silvoarable system over time may be converted to a silvopastoral system**

A Silvoarable system   **B** Silvopastoral system

Photographs: ©Paul Burgess

Individual trees on pasture and crop land

One type of agroforestry that can be silvopastoral or silvoarable is the presence of individual trees and bushes in agricultural fields. It is estimated that there are approximately 35,000 ha of single trees and bushes in the UK[10].

> Chapter 1
> What is Agroforestry?

Hedgerows, shelterbelts and riparian buffer strips

Hedgerows, shelterbelts and riparian buffer strips are probably the most visible and recognisable forms of agroforestry in the UK. Lawson et al. refers to them as "linear" forms of agroforestry where trees are grown between parcels of land (Figure 7).

Figure 7: Agroforestry between fields includes

A Hedgerows
Hedgerow system in North Devon

B Tree-line hedgerows
Tree-line hedgerow in Southern England

C Shelterbelts
Italian poplar shelterbelt at East Malling in Kent

D Riparian buffer strips
Riparian buffer strip on the River Wye bordering Herefordshire and Gloucestershire

Photographs: ©Paul Burgess

Forest farming

One definition of forest farming is using forested areas to harvest naturally occurring speciality crops. Martin Crawford of the Agroforestry Research Trust has identified food, decorative and handicraft products, mulches and botanicals as possible outputs from a forest farming system[11]. To our knowledge there is no definitive study of the extent and value of forest farming in the UK, however it is potentially significant. The Ministerial Conference on the Protection of Forests in Europe (2015)[12] estimated that the annual total value of plant and fungal products harvested from woodlands across Europe was £1,197 million (1,680 million Euros), with a further £202 million (283 million Euros) from wild honey and bees wax.

Homegardens

Multi-layers of vegetation (often referred to as homegardens or kitchen gardens) are typically in urban areas or on smallholdings and can supply fruits and vegetables at an individual level. Across Europe, the LUCAS 2012 database indicates that they occupy 1.8 million ha of land in Europe (8.3% of all land occupied by agroforestry practices)[4]. To our knowledge, we are not aware of an assessment of the extent and value of homegardens in the UK.

Why agroforestry?

Agroforestry offers a joined-up way of thinking about rural land use that addresses the negative environmental effects of intensive agriculture and also addresses climate change.

There are clear environmental benefits from agroforestry, relative to agriculture alone or forestry alone, including increased biodiversity, reduced run-off, increased carbon sequestration and reduced water pollution[13]. The most appropriate form of agroforestry will depend on the individual farm situation but there are forms of agroforestry for arable farmers, livestock farmers, horticulturalists, foresters and householders. This handbook explains how you can maximise the benefits and minimise the disadvantages when you are considering, designing or implementing an agroforestry system.

Chapter 2
Agroforestry systems design
Prof. Steven M. Newman, BioDiversity International Ltd

Introduction and aims

Chapter 1 outlined the broad classes of agroforestry systems based on the components of the system. The overall aim of this chapter is to provide a framework for agroforestry design. The specific aims are to show the special opportunities and challenges of agroforestry and working with trees; provide practical advice on developing systems based on different starting points in terms of land use; and consider different scales of operation.

©Jo Smith, Organic Research Centre

Chapter 2
Agroforestry systems design

Introduction to agroforestry design

If you talk to a landowner in the UK about agroforestry, the two most common comments tend to be: "I already have trees on my land so what is new?" and "Tree planting on my land may benefit the next generation but I cannot see it being profitable in the short term".

However, agroforestry uses the latest insights from agroecology that show how woody plants can improve land use efficiency. Also, woody plants can be used to produce a far greater range of products and services to society in the UK than we currently recognise and make use of. Climate change will only increase this demand.

There are other people interested in what you are doing on your farm because of the wider benefits that agroforestry can play. Trees also play an important role in hydrological functions. As well as helping to manage wet areas on farms, farm trees can also play an important role in controlling both the quantity and quality of water moving across the landscape. Even relatively young trees can significantly increase rates of water infiltration. Two-year-old blackthorn planted in silvopastures at Pontbren in Wales had infiltration rates 60 times higher than the surrounding pasture and these increased as the trees aged[1].

To capitalise fully on these opportunities, we need to use new varieties and forms of woody plants and manage them in new ways.

In terms of quick returns from woody plants, it is important to recognise that whilst 20th-century 'forestry' focused on species where it took decades to get a financial return, 21st-century agroforestry will be looking to get stable, socially stable and/or climate-smart financial returns in less than five years.

The two main challenges for agroforestry in the UK are: firstly, some of the best practice can only be seen in other countries; secondly, until recently, UK policy did not support agroforestry. For example, arable farmers could lose area payments if they planted trees on arable land.

Chapter 2
Agroforestry systems design

The key elements of agroforestry design

As described in Chapter 1, agroforestry management seeks to derive benefits from the ecological and economic interactions between trees and farming. An important focus of agroforestry design is to manage the tree, crop and/or livestock components in ways to optimise the ecological and/or economic benefits.

A useful acronym to help your design is **PAMASAL**.

Purpose:
Where do you want to get to?

Advice:
Where will you get advice and support?

Measures of success:
How will you measure efficiency, effectiveness and impact?

Agroecology:
How will you capitalise on agroecology or in other words let nature do some of the work?

Starting points:
Where do you want to start from? E.g. pasture land, arable land, or an orchard etc.

Adaptive management:
As things develop, what tools will help you to carry out adaptive management?

Layout:
What varieties/species, spatial arrangement and sequencing will be part of your design?

Chapter 2
Agroforestry systems design

The Purpose of your agroforestry project or intervention

Any form of diversification is a challenge. Managing and optimising more than one thing is not easy, therefore it is important to have clarity of purpose with a clear end point in mind. Have a clear date of attainment and an idea of the beneficiary. Is it yourself, a client or a member of your family for instance? If it is to form part of an ultimate inheritance there may be tax implications.

Table 2: **Examples of purposes of an agroforestry project with a potential role of the woody component**

Purpose of the agroforestry project	Possible role of the woody component
1. Increase profit but keep growing crops on my large farm	Increase the yield of the crop and/or provide an additional woody crop
2. Increase the income from my woodlands and develop a medicinal plant business	Act as shade for a medicinal plant where shading improves quality and hence price
3. Increase the income from my orchards and provide a more social benefit	Trees provide a pleasant environment for camping or glamping Can provide a healthy food option for added value products
4. Make my hedgerows a profit centre rather than a cost centre without major capital cost	Hedges produce fruits and nuts that can be harvested at low cost for a premium market with a net positive gross margin
5. Maintain the same level of food productivity on my land and produce energy for export	Woody biomass is an energy feedstock and is planted in a way that does not reduce crop or animal productivity
6. Maintain profit on my land and reduce the need for labour	Woody component serves to improve return per unit labour

Chapter 2
Agroforestry systems design

Advice

It should now be clear from Table 2 that agroforestry is more than just forestry on agricultural land or doing agriculture in a forest or woodland. It can involve high value horticulture, energy cropping and animal nutrition.

To get a rapid income will require products of high value with specialised markets. Reading around the subject gives basic knowledge but best practice would be to obtain advice on issues like the suitability of the site, prescriptions for management, harvesting, processing and sales.

Consider forming a partnership where the woody component and the non-woody component are managed by separate partners. Examples include i) arrangements between an orchard owner and sheep owner, and ii) the Dartington Hall silvoarable agroforestry system, see https://www.dartington.org/trust-test-new-multi-agency-agroforestry-model-48-acres/

Measures of success

Many equate success to financial gain, however there are other important factors to consider when measuring the success of an agroforestry system. These include efficiency, effectiveness and impact. Also important is that many of the benefits from agroforestry adoption emerge as things develop, they often are not part of the initial plans. Which is one reason why adaptive management is so crucial.

Efficiency

At its simplest efficiency is a ratio of a physical output divided by a physical input. However there are various ways of looking at this ratio.

Agronomic: Farmers and advisors in the UK are very familiar with yield per unit area or live weight gain per unit area. UK foresters are very familiar with timber volumes per unit area to be expected from a certain quality of site (yield class). Agroforestry can include these measures but may also need to get over an initial view that managing systems with multiple outputs is more trouble than they are worth.

The abbreviation M.F.M. is useful when looking at yield per unit area, when testing whether combining more than one activity on the same unit of land is worthwhile. The abbreviation consists of three completely different measures that an agroforestry designer can measure and optimise: Main crop, Feedstock and Multicrop yields.

Main crop yield: Where the grower knows what their main crop will be, e.g. winter wheat, spring lambs or number of cricket bat willow trees. Here the test is that trees can be added to agricultural ventures or agricultural ventures can be incorporated to tree production systems only if the yield of the main crop is increased by the addition or not affected at all. For example, trees can increase wheat yield in some situations through windbreak effects. In the UK buyers of willow for cricket bats are happy to integrate this with grazing if the trees are protected.

Feedstock yield: Where the total biomass per unit area is of interest. Individual species are less important than the physical and chemical nature of the feedstock. Examples of a feedstock include energy crops, biomass, fodder or mixed 'grain' for breadmaking. Here the comparison is the maximum yield attained by a sole crop compared to the combined yield of the mixture. Table 3 gives some hypothetical examples of successful designs.

Table 3: **Hypothetical examples of successful designs of feedstock agroforestry**

Type of 'feedstock'	Yield — Herbaceous component grown alone	Yield — Woody component grown alone	Total from woody herbaceous mixture	Comment
Biomass	5 tonnes per ha from wheat straw	10 tonnes per ha from willow	12 tonnes per ha	Straw and wood pellet mixture had desirable combustion properties
Animal feed	10 tonnes per ha from fodder radish	5 tonnes per ha from alder tree leaves	14 tonnes per ha	Mixture had desirable feed characteristics and did not reduce milk yield per cow or live weight gain
Bread making 'grains'	10 tonnes per ha from wheat	3 tonnes per ha from sweet chestnut	11 tonnes per ha	Mixture produced acceptable loaves as far as the consumer group was concerned

Chapter 2
Agroforestry systems design

Multicrop yield: Is it more agronomically efficient to combine two or more entities on the same piece of land?

In formal terms this calculation is known as the Land Equivalent Ratio (LER).

$$LER = \frac{\text{Combined Yield / Solo Yield crop A}}{100/100} + \frac{\text{Combined Yield / Solo Yield crop B}}{100/100} = 2$$

2 is a target and has so far only been found for mixtures of pear and radish[2].

If crop A is walnut and crop B is wheat you might get 80% target yield of wheat and 40% target yield of walnut from the same hectare.

This would give 40/100 + 80/100 = 1.2

1.2 means that there is a 20% yield advantage or, put in another way, 20% more land would be required to obtain the same yield from monocultures.

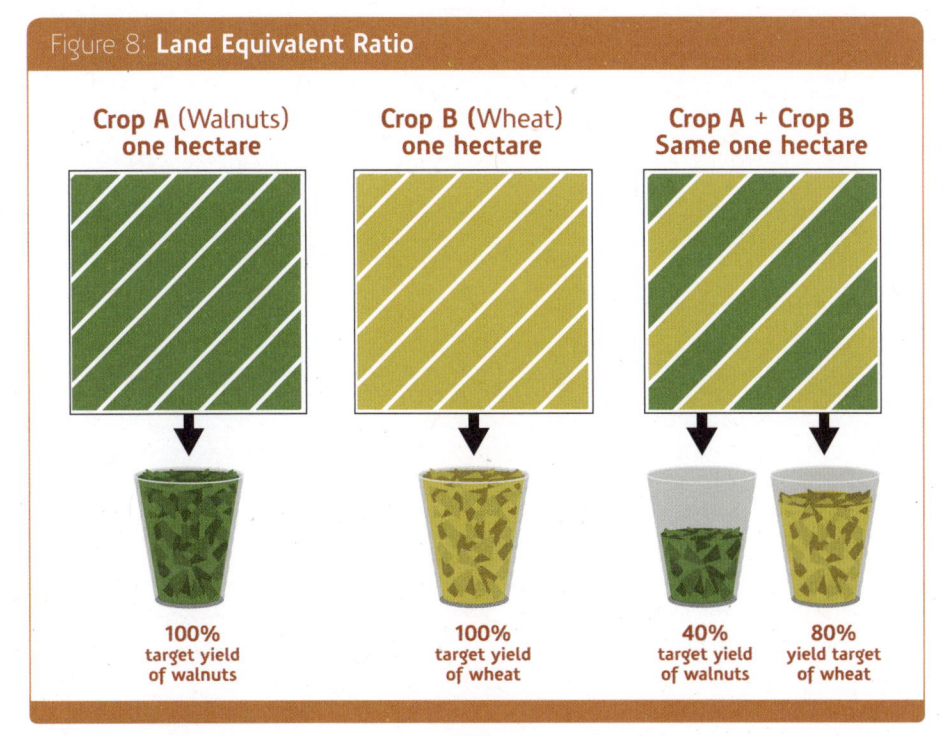

Figure 8: **Land Equivalent Ratio**

Crop A (Walnuts) one hectare — 100% target yield of walnuts

Crop B (Wheat) one hectare — 100% target yield of wheat

Crop A + Crop B Same one hectare — 40% target yield of walnuts | 80% yield target of wheat

Profitability: The standard approach for looking at the profitability of enterprises, e.g. crop, livestock or forestry, on agricultural land is to use a gross margin approach for enterprises and to calculate the fixed cost of running the farm or estate as a whole. A useful reference is the John Nix Pocketbook for Farm Management[3]. Agroforestry approaches will include looking at the opportunity cost of time, i.e. it may be several years before a sale is made. At the design stage clarify the following questions:

1. Is annual profit the key measure or return on capital?
2. How important are assets and asset securitisation?
3. Are discounted cash-flow approaches relevant if you see the future income from woody plants as a tax-free gift to the next generation?

Other ways of measuring efficiency could include:

Yield per unit management labour: This is very important if labour availability is going to be limited in the future, e.g. if family members die or leave farming. There are also opportunities to even out labour demands through different seasons using tree crops.

Yield per unit artificial fertiliser: Important if fertiliser is a major part of variable costs or if carbon footprint is to be reduced.

Yield per unit water: This may be important if irrigation is a major part of variable costs.

Effectiveness and impact

If you would like financial support from the government or charitable sources for your agroforestry design, then it would be useful to show how impact (spread) might be achieved and simply monitored. By designing the agroforestry carefully with impact pathways in mind, powerful lessons will be learned in a short space of time. In most cases impact is greater than envisaged in a design due to unforeseen impact pathways.

Effectiveness: This can be defined as the relationship between activities and an 'outcome'. For example, if the desired outcome is 'people adopt behaviours with a lower carbon footprint' then one could compare behaviours attributed to agroforestry farms with those on high-input arable farms using a system boundary of a county.

Impact: This is the spread of effects outside the management system boundary, e.g. the agroforestry monitor farm led to changes in policy or procedures that affected the whole country. This could be linked to a cost-benefit analysis of societal benefits or a financial analysis of increased profit to a company if taxation reform led to more sales of a particular product.

Agroecology

The carbon footprint and environmental impact of most agricultural inputs is unacceptably high given that there are alternatives that could become carbon negative.

The agroecological paradigm for the 21st century seeks to avoid these undesirable effects and contrives to let nature do the work of fertilising crops, feeding animals and reducing pest and disease problems. It recognises that increased woodiness and hence agroforestry is a key pathway in the evolution of successful agroecological approaches.

Here are some ways of assessing the agroecological benefits that agroforestry can bring.

Modulation: Where one component manages or modulates the **physical** environment of another component. A tree may reduce the heat loss from livestock through acting as a windbreak thus giving increased profit through better live weight gain per unit feed.

Synthesis: Where one component changes the **chemical** environment of another component. A shrub may fix atmospheric nitrogen and this nitrogen may be made available to an adjacent non-nitrogen fixing crop when the leaves fall in the autumn or when the shrub is coppiced or pollarded (nitrogen is derived from the above-ground biomass and the increased root death caused by reduced above-ground biomass).

Partitioning: This is also known as **sharing**. It can happen in both space and time. A deep-rooted tree may get its phosphate requirements from a deep layer of soil below the roots of an adjacent crop and avoid competition. On a small plot a tall tree uses the light that would fall on areas outside the plot boundary. Late leafing trees may intercept light after the removal of a crop, e.g. a cereal that has started grain filling (doesn't need light for photosynthesis) after May.

Chapter 2
Agroforestry systems design

Starting points

Any design starts with the land. This will have properties linked to soil, elevation and exposure that will determine species selection. There is some guidance to help with this, such as the Forestry Commission's Ecological Site Classification Decision Support System (ESC-DSS)[4]. Take time to review and observe your current land use. You are likely to be starting with one or more of the following: pasture, arable, orchard, woodland, or 'edge land'.

Radical change is not necessarily the most appropriate; you may be able to gradually adapt your existing systems. For instance, planting trees into your pasture land, or introducing animals into orchards.

Consider what size and type of tree to plant. For instance, bare-root or container-grown, whips, feathers or standards. Large-sized stock may be the only option for fruit trees, but be aware that it may need more initial maintenance. In summer 2018 planted stock really suffered from water stress, but smaller-sized whips were generally less affected.

It is worth noting that silvopastoral systems can be changed to silvoarable systems and vice versa. A key example is the traditional grazed cherry orchards found in Kent and elsewhere before 1940. Hoare (1928)[5] outlines that these started off as arable fields and the trees once established were intercropped with vegetables for the first three to five years. The sward was then established and livestock were admitted.

Orchard systems of agroforestry can incorporate livestock or intercropping and may produce timber as a valuable by-product. Much of the walnut veneer for luxury cars and furniture now comes from old orchards.

Woodland: if woodland is already managed for a pheasant shoot, then this is already a form of agroforestry. The management of the trees and the understorey vegetation has a great influence on the feed availability and holding ability of the woodland.

Markets

Before diversifying, consider the market for a specific tree variety and research the demand for a particular product. Specialist knowledge may be required for specific markets which will need to be taken into account, as well as factors such as cost and management.

The costs of plants, establishment and management can be expensive. Will you go for a multipurpose tree, e.g. walnut for nuts, nut oil, and timber (saw log or veneer) or will you grow one specialised product, e.g. green nuts for pickling? For timber do you want to produce homogeneous clean pruned straight 'telegraph pole' form or get higher returns from specialised markets, e.g. bent timbers for half-timbered housing, splayed roots for hurley sticks, or butt logs from pollarded trees? Will campers or glampers pay more for a closed canopy or a parkland experience? If the main market is payment for ecosystem services such as flood control, spatial layout and tree management will be of central importance.

Table 4 provides examples of what tree to consider for each of the main markets in the UK. Remember though that many trees will have multiple potential uses. For more detail on tree species selection, see page 139.

Table 4: **Trees to consider for UK market opportunities**

Market	Tree to consider
Woodfuel	Oak, beech, hazel
Specialty timber	Cricket bat willow, walnut for furniture
Biomass	Hybrid willow or poplar
Bedding	Pine, spruce
Fodder	Elm, willow, poplar
Fruit	Apple, pear, cherry
Nuts	Walnut, hazel, chestnut
Herbs	Elder, hawthorn, ginkgo
Woodcrete	Cedar, oak, beech

Chapter 2
Agroforestry systems design

Adaptive management

Over time as the trees and woody plants grow, the effects on adjacent land use will change, and new results, opportunities and constraints will appear. It may be useful to design an agroforestry management plan that uses phases such as 'establishment', 'mature' and 'over mature'. Trees at establishment need protection from livestock and/or weed growth. During development they will need to be managed and maintained to ensure they deliver the right product. At the mature phase they may be yielding valuable nuts, fodder and fruit so may need nutrients and pruning. Also at the mature phase, plans will be required for removal or if they are to be left in situ, options for rewilding or biodiversity enhancement.

Trees at maturity can present a special environment that lends itself to sporting pursuits and amenity. A mature silvopastoral system or parkland may add value to a property when it comes up for sale.

Protection

Trees in agroforestry systems are susceptible to damage from pests and livestock, particularly in silvopastoral systems. Protection is vital and expensive and requires a management plan for protection.

There is a wide range of tree shelters. It has been found that the solid tree guards that tend to be used in woodland/forest planting are not so suitable for widely spaced trees – experience in France has shown that mesh shelters are often better.

Inspect continually as sheep will use posts for rubbing and can push over guards. In dry spells clay ground can crack and loosen posts as well. Tree protection needs to allow access for maintenance – sheep and cattle guards may restrict access for pruning and opening and closing guards takes time.

Figure 9: **Tree protection**

Mesh deer guards on perry pears at Eastbrook Farm, Wiltshire

©Ben Raskin

Table 5 has some examples of pests you many encounter and how to potentially protect from them. Often you will have to weigh up the risk of damage against the cost of protection.

Table 5: Examples of types of tree protection

Pest	Protection
Deer	Tree guards, deer fencing around the whole field, shooting, electric fencing
Rabbit and hare	Fencing, tree guards, shooting
Squirrel	Shooting, keeping open areas between rows of trees – silvoarable systems might be better for this. New techniques such as electric fencing systems, contraception and bolt traps are in development
Voles	Vole tree guards. Keeping grass short around trees also helps
Cow	Very solid fencing, electric fences
Sheep	Solid fencing, electric fencing
Chicken	Tree guard to prevent scratching right up to the trunk
Bird pests	Pigeons or other birds can damage trees by landing on them. In areas with few trees consider putting larger roosting cane for each tree

Mulching

Mulch sheets are commonly used in horticulture, e.g. for the production of strawberries. In agroforestry they also have a role in the management of the tree component in order to reduce interference from weeds or to improve soil conditions. The first design variable concerns the choice of living or dead mulches. Dead mulches include the use of plastic strips or disks into which the trees can be planted at establishment. Natural alternatives such as woodchip, sheeps wool, cardboard and fabric mulches like hemp can also be used. Key design features to consider are the level of biodegradability and permeability to rainfall. Living mulches include the planting of trees into strips of wildflower mixes or mixes of cover crops. The latter often includes nitrogen-fixing species as part of the maintenance of soil fertility. Further information on mulches is given in the chapters dealing with silvoarable systems and silvopastoral systems.

Layout

It is easy to think about the layout of a row crop in a field. The distance from the edge is set and a precision drill does the rest as the coulters will give the inter-row distance. For agroforestry design, things are a little more complicated. It can be useful to think about cropping in a cube or in other words, a three-dimensional space over a set multi-year time frame. Here are some overarching factors that determine the layout.

Ergonomic factors: Agroforestry system layouts need to be ergonomically designed to ensure access of machinery and consideration of safety. Felling and harvesting large trees can be hazardous and it is important that large equipment can easily get to the trees and take any logs to a roadside or ride. Most managers will have standard farm equipment and it will be cheaper in the long run to adapt the layout to the size of the machinery available rather than buy specialised equipment. Many grazing licences in the UK have a clause that makes removal of thistles by the grazier essential. This is often done by mowing. In silvopastoral systems the trees should be planted in a way that gives easy access to the mower. The distance between the trees should be a multiple of the mower width plus approximately 10%. For rows of trees it is best that these are arranged to be parallel with the longest axis of the field.

Agroecological factors: Agroecological considerations also affect the layout design. For instance to modulate wind, plant trees in a solid row. The aspect of the rows will affect the distribution of light on the ground and any self-shading of the trees. North–south orientation of the trees suits a system where apples are the main crop. If the understorey has an equivalent value, then other orientations may be optimal. You might also need to consider prevailing winds or air currents.

For instance some trees such as walnut are said to produce volatile compounds that can mediate insect or disease populations. Ensuring the wind blows those compounds in the right direction is crucial when designing the spatial layout. Slopes will also have an impact.

Aesthetic factors: As trees grow they soon become conspicuous in the landscape. Consider visual amenity and aesthetics in the design. For instance, a block of trees on a hillside may look more attractive if the edges are wavy and follow the contours rather than present a hard, straight edge.

Chapter 2
Agroforestry systems design

> **The four variables that affect layout are:**
> - **Species**
> - **Spatial pattern on the ground**
> - **Use or occupation of three-dimensional space**
> - **Phasing**

Species and varieties

There are no limits on what species can be brought together in agroforestry if the site is suitable and mitigating measures are made, however the following guiding thoughts may be useful:

- Species may harbour pests or disease which could infect an adjacent species, e.g. rhododendron fungal diseases can affect larch
- Trees with late leafing and or short leaf area duration may be optimal in some systems
- Variety is critical for fruit and nut trees. Grafting can dramatically improve precocity (age of bearing)
- Different breeds of sheep have different bark-stripping behaviours

Spatial pattern on the ground

Trees could be planted in five basic patterns:

| Blocks with straight edges (normally more than nine trees). | Groups with round edges (normally clumps of five or more). | Borders with trees planted around the edge of the field (can be in single or multiple rows). | Strips in the middle of the field (can be in single or multiple rows). | Individual (single or more as in a parkland landscape). |

Use or occupation of three-dimensional space

Systems can be multi-layered with an understorey herbaceous layer, shrub layers and tree layers. This is often the case in the form of agroforestry known as forest gardens[6].

It is possible to prune trees by removing the lower branches and essentially lift the canopy. This can be useful for ergonomic reasons and in most cases removal of less than 30% of the canopy in this way will not affect growth.

Phasing

This concerns the sequence of events on the ground in relation to the developmental phase of the woody component. Some examples include:

Catch crop: This is where cropping or grazing happens at the establishment phase of the trees. The practice may be stopped at canopy closure or some other developmental stage.

Relay: This could be as in the case of short (three-year cycle) rotation coppice in a silvoarable system outlined here:

- Year 1 coppice established
- Year 2–4 intercrop taken
- Year 5 canopy closes, and row coppiced
- Year 6–8 intercrop taken
- Year 9 canopy closes, and row coppiced
- etc.

Phasing within the year can also be possible in the case of deciduous trees with different intercrops being used in full-leaf and leafless phases.

Depending on the system you are using (for example in a clump system or a row system) you may want to grow high-value trees with nurse plants and shrubs to help with their form (and provide additional benefits such as shelter).

Chapter 2
Agroforestry systems design

Practical considerations when designing specific agroforestry systems in the UK

Introduction

This manual is a short guide to practical agroforestry in the UK. Other manuals and guidelines on the general and species-specific aspects of agriculture, horticulture and forestry are available, therefore they will not be repeated here.

The general variables dealt with by these manuals include: land suitability and preparation, establishment, crop/tree management, livestock management, harvesting, markets, grants and taxation.

The practical examples that follow deal with a range of observations and reflections created by taking the agroforestry compared to conventional individual enterprise option. They are based on 35 years of pilot collaborative trials with farmers and landowners. It was understood that mistakes would be inevitable and useful.

In addition, there is now a greater potential for designing agroforestry systems at the landscape scale rather than the field/woodland or plot scale.

Shropshire sheep grazing an orchard

Chapter 2
Agroforestry systems design

Practical design example silvopastoral

To solve the problem of agricultural surplus, e.g. butter, the then Ministry of Agriculture, Fisheries and Food (MAFF) supported experimental trials across the country linked to an economic analysis exercise known as bioeconomic simulation modelling[7]. These trials used variants of the model planting protocol of 100, 200 and 400 stems per ha with forestry controls at 2,500 stems per hectare. The idea was to grow timber in pasture as part of a transitional land use moving from livestock production to forestry, since farmers and landowners would not be willing to replace grasslands with forest plantations overnight.

The reflections outlined here come from a trial with blocks of ash planted on permanent pasture in Buckinghamshire in 1986. Sheep were introduced as part of a rotational grazing system.

Surprises

- Sheep liked the ash foliage and bark due to a potential nutritional or medicinal property, causing severe damage to the trees. Individual tree protection of a tree stake and netlon to a minimum height of 1.2 m was costly. Constant inspection was required due to damage from sheep. The team trialled a product called WOBRA, a non-toxic, food-grade standard, sand-based product, applied by brush to the tree trunks. This acted as a very powerful deterrent and there was no more damage from the flock or subsequent for over 30 years from one application.
- After approximately 10 years the 400 stem per ha canopy closed, and the pasture began to fail. Sheep were attracted to the plot and their 'camping' meant that soil compaction was greater.
- The pasture in the 100 stem per ha plot has still not been dramatically affected to this day, by either the sheep and the cattle introduced some 32 years later.
- Ash is leafless for a good part of the year and therefore not taking up water and nutrients.
- Thistle control was a challenge during the early stages of the 5 m plot proving difficult with a tractor drawn mower. In the 10 m plot there was no problem.
- There is a strong variation in tree form and understorey in the different plots given they are the same age.
- The best and most profitable produce was potentially wood for hurley sticks.

Complexities

Ruminant health and nutrition is complex. We are only now rediscovering the value of tree fodder in the UK. This can arise as part of tree form management and could be a major by-product. Tannins can have a positive effect on live weight gain by affecting gut microflora.

Lessons learned

It may be more profitable to develop a sheep silvopastoral system by starting with a silvoarable system or at least trees with a fodder crop to be cut and not grazed due to the high cost of individual tree protection.

If you are growing crops for fodder, how will animals use that? Will they be allowed direct access or will the fodder be cut and stored? Different systems can take vastly different amounts of time (and cost) and need to be understood before any planting is done.

Deciduous trees are very useful in the case of design linked to temporal partitioning. Mulberry would have an even shorter leaf area duration than ash and most fruit trees including those for cider and perry could be part of a mixed silvopastoral systems approach.

The landowner gets a great benefit from a 100 stem per ha system. They get full rent from a grazier and a bonus of a tree crop and or final timber. The grazier is a key component and needs to be consulted in relation to ergonomic issues, e.g. spacing so that mechanised thistle mowing can take place.

Rules of thumb

100 stem per ha systems look like a suitable starting point for silvopastoral system designs where production from the understorey is important long term.

Silvopastoral systems should be considered within any plan for rewilding. They can provide a more profitable alternative in the transition period when no income is expected. They also offer the opportunity for food production and biodiversity on the same area of land.

Chapter 2
Agroforestry systems design

Practical design example silvoarable

In the silvoarable systems set up in the UK in the late 1980s by Bryant and May arable farmers grew poplar on at least a 22-year rotation to produce peeler logs. From this system we learnt that:

- Greater distance was needed between tree rows to allow for bigger farm machinery
- Weeds were prevented from spreading to the arable crop by installing black plastic mulch strips prior to planting the trees cuttings (rods)
- Poplar was in demand for an emerging market for energy from biomass. Trees could be planted at close spacing in the rows (1 m) and could be harvested (coppiced) on a short rotation of three to five years
- Better clones of poplar were developed, e.g. from Belgium for disease resistance and fast growth.

More details of the work of the silvoarable teams in the UK can be found in Burgess et al. 2005[8] and Burgess et al. (2003)[9].

Surprises

- It was easy to lay the mulch and plant the system only taking two people one day to set up the 10-acre trial.
- Over 99% of the cuttings took and grew at an astonishing rate.
- The plastic mulch withstood the weather and agricultural operations and didn't have a negative environmental impact as expected. A considerable amount of leaf litter rapidly covered the mulch making it invisible in the second year. The mulch contributed significantly to the tree growth rate by elevating the soil temperature in the top layer by 1°C per day on average. However, the removal of the black plastic sheet can be very laborious if problems occur. In this trial there was never a need to do this.
- The mulch had a positive effect on increasing biodiversity by creating an important habitat for small mammals such as voles as well as slow worms and snake species. Annual ploughing did not seem to damage tree roots.
- Very few farms have managed to export energy linked to the judicious use of farm waste including woody waste and chips by using mini CHP approaches. Although the system was focused on the production of

peeler logs, the current UK market for such logs is minimal. Hence there has been an increasing tendency for new silvoarable systems to include either a mix of hardwood timber species, or fruit or nut trees.

- Both the farmers and the researchers thought that the tree roots would grow like they do in a forest (radially symmetrical and mimic the canopy), that ploughing would damage them and check the tree growth and that the crop's growth would be affected in a linear way reducing yield annually. None of this turned out to be the case. The tree roots behaved in two ways. Some roots became rope-like and travelled parallel to the edge of the mulch strip. Adventitious roots spread out and under the crop row. The plough did very little damage. The crop yield in in the alleys was not significantly affected for the first five years.

Complexities

Trees over 2 m tall created a windbreak or shelter effect. Sampling in the middle zone of the crop strip showed over yielding compared to the control, though this is not generally the case and this could eventually compensate for the loss of land due to the presence of the mulch strip. Tree ideotypes can be selected to give less shade and to develop crops that are shade tolerant.

Weed growth from the tree row into the cropped alleys can also be an issue.

Lessons learned

Trees respond more to the subsoil than agricultural crops and crop growth is not a reliable indicator for potential tree growth.

Rules of thumb

For arable crops there may be a severe reduction of yield if crops are planted between dense plantings of vigorous trees when the tree height becomes equivalent to the inter-row distance.

Plant a mixture. Monocultures will always be risky. Climate change appears to be increasing tree disease and insect problems in the UK.

Practical design example walnut trial

One of the most significant agroforestry trials in the UK was started in the late 1980s by a team from the Open University using grafted Persian walnut trees *Juglans regia*[10, 11]. The trials were established in Buckinghamshire (silvoarable) and Essex (silvopastoral).

Surprises

- It is easy to grow walnut for nuts in the UK if the tree has adequate shelter from wind. The main problem is control of squirrels if growing for table nuts. In reflection a better design for the system would be to have wider arable alleys to allow for modern arable machinery and to prune and or pleach the trees so that they form a continuous hedge at a height of less than 2 m. This would facilitate picking and may increase the yield on tip-bearing varieties. It would also mean easier access for the arable machinery.
- A surprising number of products can be obtained from walnut trees including wines, dyes, abrasives, oils, saw logs and veneer.

Figure 10: **Walnut for agroforestry**

Walnuts and apples planted in the lambing field

©Mark Measures, Organic Research Centre

Complexities

With so many potential products, walnut orchard agroforestry design can be very complicated. Table 6 illustrates some aspects of this with comments.

Table 6: Some of the complexities of walnut orchard agroforestry		
Form of walnut	**Product**	**Comment**
Black walnut from seed	Timber and nuts from a large timber tree	Further work is needed on seed provenance for the UK linked to timber production. Nuts are an occasional by-product. UK consumers are not used to the hard shell and strong flavour
Hybrid black and Persian walnut	Timber and nuts from a large timber tree	Has hybrid vigour and could be a replacement for ash. Excellent saw logs. Nuts are an occasional by-product
Grafted Persian walnut	Leaves and nuts. Small tree if pickling nuts are the main product	All of the non-timber products have a myriad of uses. Graft union may create burrs for valuable veneer market
Top-worked Persian walnut	Quality nuts and quality timber from an intermediate size tree	The idea is that the 'rootstock' is managed as a quality butt log. Once the tree develops to the requisite size it can then be top worked by grafting scion wood from superior nut varieties

There is a great potential to develop nut agroforestry in the UK as a 'climate smart' and nutrition-sensitive alternative to many UK food production system practices. Global markets for nuts and nut products are still increasing at an astonishing rate as China and India increase demand.

Rules of thumb

Given the complexity of orchard agroforestry, partnerships with companies that are concerned with sustainable procurement are the recommended option.

Harvesting/managing the tree crop should be one of the major design drivers of spatial layout and tree form.

Chapter 2
Agroforestry systems design

Woodland/forest systems including forest gardens

There are two interesting systems in the UK. The first is intercropping woodland with high-value plants/fungi and the second is creating forest gardens. As yet there are very few commercial examples of either in the UK. Forest gardening now has a major following and there are several manuals available.

Challenges

The major challenge for woodland intercropping relates to conservation concerns of introducing non-native plants into UK systems of high conservation value.

The main challenges for forest garden agroforestry appear to be: **1** to improve profitability of the productive component; and **2** to provide adequate levels of caloric staples from lower storeys of production. Forest gardens can be made profitable as visitor attractions with linked training and plant sales ventures, but sales of forest garden produce appear to have been limited so far.

Most forest gardens, as currently designed, rapidly develop into a closed canopy. This gives rise to shade levels that would adversely affect the yield of crops like potato, which might be an obvious choice. Shade-tolerant alternatives could be a solution, in this instance crops such as yams and yacon.

Rules of thumb

Two major rules of thumb that are emerging from forest garden experience are: **1** use wider spacing so that parts of the garden never attain 100% canopy cover; and **2** consider temporal partitioning when designing the layers of the system. This could lead to opportunities for higher productivity and species diversity.

Hedgerow/buffer strip systems

Chapter 5 of this manual is devoted exclusively to this topic. The major considerations in terms of design will be linked to creating good conditions for tree establishment, aspect effects and harvesting. In some cases, there may be jurisdiction issues concerning responsibilities at different distances from the edges of roads, waterways and railways.

Landscape, estates and partnerships

With changes in agricultural subsidies moving towards payments for results, it is clear that there is now great potential for agroforestry designs to be considered at the landscape and especially watershed scale. Agroforestry at the landscape scale can be considered as a continuum from single trees to parklands to row, or hedgerow agroforestry to woodland/forest blocks.

Agroforestry design at the estate level could consider the potential for linking agroforestry designs to the provision of low-cost housing in a manner that is carbon negative and can restore community and nature within the countryside. See Newman (2018)[12].

Partnerships are very important when designing agroforestry at a large scale. One of the most promising options globally is the concept of tripartite environmental stewardship contracts, where the tree partners include a landowner, local people with an interest in sustainable rural livelihoods, and an entrepreneur or broker who can get the best price for an environmental product or ecosystem service. The three parties decide on the share of equity and responsibilities and set indicators of achievement for the contract period which is normally greater than 10 years. This model could be used in urban and peri-urban agroforestry landscapes, so the local community can be involved in combined landscape management and energy cropping (Newman 1985)[13] with food as a possible by-product.

Observation on wildlife

The most surprising aspect about agroforestry (this is also true of rewilding[14]), is the rate at which key wildlife species return after conventional agriculture is held back from even the smallest sites.

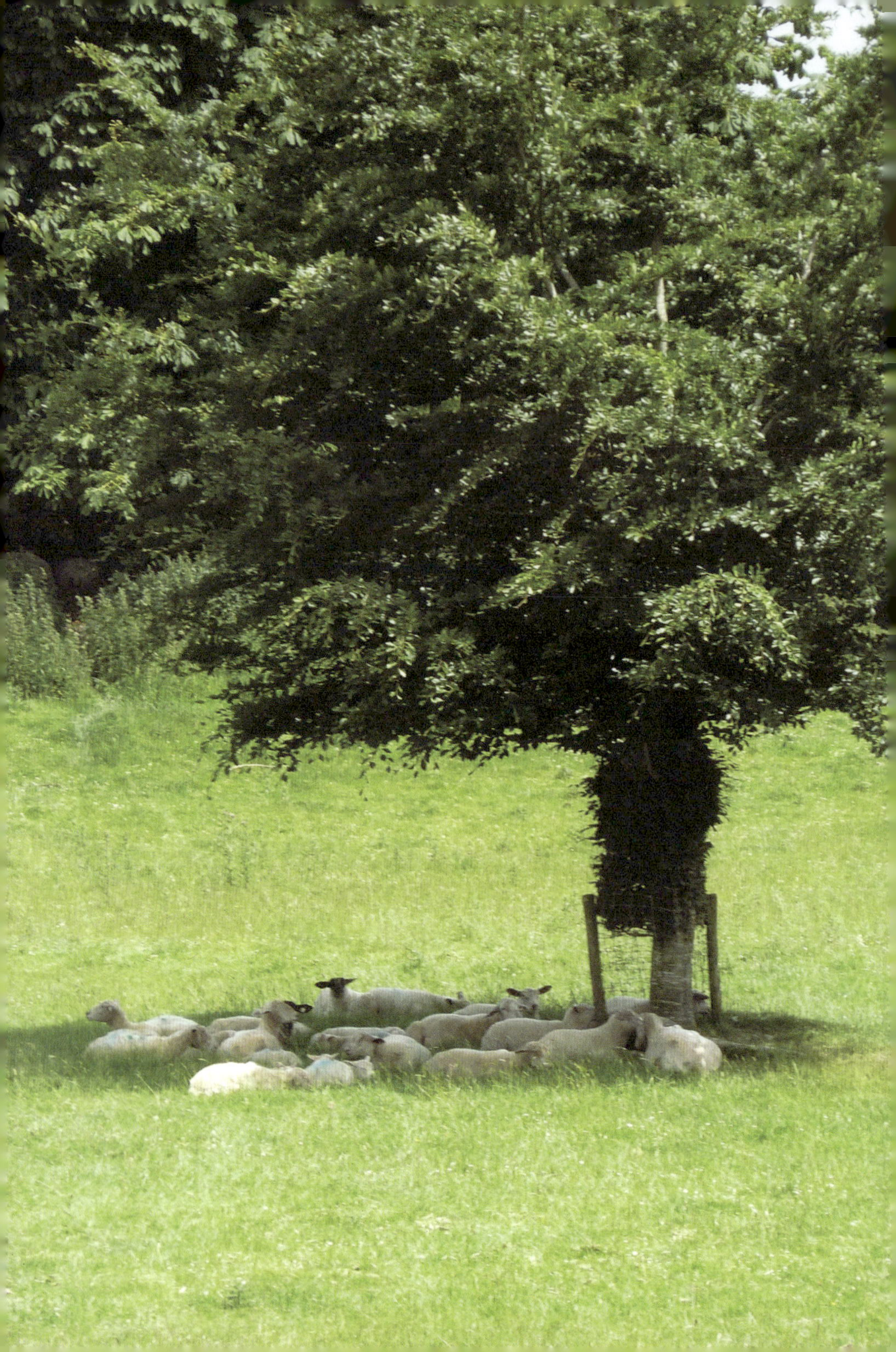

Chapter 3
Silvopasture

Dr Tim Pagella, Bangor University with
Dr Lindsay Whistance, Organic Research Centre

What is silvopasture?

Silvopasture is a management practice where trees are integrated into the same unit of land as livestock (i.e. ruminants, pigs or poultry) and where that interaction results in direct economic and/or ecological gains to the farming system.

Trees can provide economic benefits if they are managed as a second crop (either for timber, firewood or for biomass) which can be marketed or used on farm to reduce costs. Trees also produce a broad range of agroecological benefits that can contribute indirectly to the bottom line. For example, trees along a field boundary can alter their surrounding microclimate by providing shade. Access to shade reduces heat stress in ruminants and can significantly increase their productivity at a minor cost to grass growth near the tree. In addition, the same trees can help dry out wet soils to enable easier vehicle access and improve the farm's biodiversity. They can also help make the farm more resilient by reducing soil loss and buffering farms against extreme events.

The ideas associated with silvopasture can be found within many traditional land use systems across the UK. It is the most common form of agroforestry[1] with one third of tree cover in Great Britain being found on working farms. As such, many farms have existing silvopasture on them perhaps without realising it. Silvopasture can involve different combinations of trees integrated into forage systems (pasture or hay) and combined with livestock production. Similarly integrating livestock into woodland areas can benefit both the trees and the animals and is also considered a form of silvopasture.

In this chapter we will first identify the benefits that trees can provide to various types of livestock systems, and then discuss the considerations for either integrating or expanding tree cover on farms.

Photograph ©Jo Smith, Organic Research Centre

How can silvopasture benefit my farming system?

The benefits associated with silvopasture fall into two broad categories: economic benefits and agroecological benefits (both on and off farm). As discussed briefly on page 45 these are not mutually exclusive, trees can and will often provide both of these benefits, but their relative importance will vary with the management priorities of the farm and the context in which the farm is found.

Direct economic benefits

As well as providing direct income streams, the provision of an additional tree crop can be a primary objective for adoption of silvopasture (as a diversification strategy) or it can be a significant side benefit where trees have been integrated to provide other benefits, such as shelter.

Maximising the economic value of trees does require careful management (both in terms of tree selection, siting and caring for the trees) to maximise the return on investment and generally also involves a longer time frame before these benefits are realised as the timber trees would need to mature before harvesting, for example. The farming context is also an important factor. More sheltered lowland silvopasture systems sited on better soils will be able to produce higher-quality timber or fruit more quickly than exposed upland silvopastoral systems.

Examples of economic benefits are discussed in Chapter 6.

Agroecological benefits

Trees provide a broad range of agroecological benefits. These benefits are often subtler (in that it is harder to put direct economic values on them) but are often critically important for the long-term sustainability of farming systems. The body of evidence associated with the agroecological benefits is growing rapidly and includes details on benefits to farm productivity – an overview of the typical benefits is provided here but the exact mixture of benefits will vary with farm context and farm objectives.

Increases to soil health

Trees help to maintain the long-term soil fertility of pastures. Trees capture nutrients leached below the grass rooting zone and return them to surface soil via litter and root turnover. Trees improve the soil's holding capacity for water and nutrients. Trees can limit compaction by animals (poaching) and increase infiltration. Under elevated stress conditions (such as drought) trees invest in their mycorrhizal associations and can scavenge water and nutrients from deeper within the soil The same is true under elevated CO_2 conditions, suggesting that trees also help resilience to climate change. In addition, trees' root systems significantly reduce soil loss from erosion. Trees also encourage beneficial soil organisms. Under silvopasture systems there are significant increases in the ratios of fungi and bacteria and increased numbers of earthworms. For most systems this increase is an indicator of a healthier soil.

Trees can also be used to reduce fertiliser costs. Selecting nitrogen-fixing trees (such as alder) can lower fertiliser use[2].

Reduction of effects of wind exposure

In areas with high exposure to wind silvopastoral systems can provide substantial shelter benefits to livestock. Livestock need significantly more energy to maintain their condition in exposed conditions. Animals with shelter use less energy to maintain core body temperature than those without access to shelter, resulting in lower feed costs and higher animal welfare. These benefits can increase the profitability of the farm system. For example, sheltered areas can contribute to 17% estimated increase in dairy milk production[3]. In sheep good shelter provision can enable live weight gains as large as 10–21%.

Figure 11: **Trees for shelter**

Ewes utilising the shelter of trees at lambing

©Lindsay Whistance, Organic Research Centre

Carefully designed silvopasture systems can extend out-wintering periods and can substantially reduce livestock mortality rates at birth or in extreme weather conditions. For example, exposure (and starvation) are responsible for anywhere between 30–60% of lamb deaths. Trials conducted in south-east Australia indicate that losses of newborn lambs were reduced by 50% where there was effective shelter in place[4]. Another study in New Zealand found that wind shelter decreased twin mortality by 14–37% and overall mortality by 10%. Shelter also reduces the risk of ewe mastitis.

Reduction of heat stress

Overheating in livestock can have significant impact on livestock productivity. Heat stress contributes to decreased live weight gains (as livestock eat less), it can lower milk production and reduce breeding efficiency. Heat stress costs US dairy farmers $1.2 billion /year in reduced milk production and reduced fertility[5]. Heat stress can reduce conception rates of ewes and lowers the libido and fertility of rams. Similarly, hens show reduced feed intake and egg weight, and lowered immune system as a result of heat stress.

Figure 12: **Dairy cows making good use of available shade**

©Lindsay Whistance, Organic Research Centre

Chapter 3
Silvopasture

Seeking shade or shelter are natural and effective animal behaviours and, in silvopasture where solar radiation can be reduced by as much as 58%, skin temperature is 4°C lower than on open pasture. As a consequence, other normal behaviour patterns such as eating and resting are better maintained. In areas with limited shading opportunities livestock will tend to clump (increasing the risk of disease, soil compaction and death of vegetation), so provision of more even shade using silvopasture can reduce this effect.

Where they have access to natural shade during heat stress periods research has shown that cattle can put on >0.5kg/day[6].

Reduced incidence of pests and diseases

Planting trees in wetter areas of the farm can provide additional health benefits to livestock. These areas are often only marginally productive. Fencing them off helps with managing stock. Trees will naturally dry soils creating conditions that are less favourable for bacteria that causes foot rot or the snails that form part of the liver fluke cycle. Whilst trees can increase risk of head flies and blow flies (by offering a habitat for them), in a well-designed silvopastoral system there are also more dung beetles. These remove faeces more quickly and combined with higher predator numbers can result in fly counts that are 40% lower than on open pasture. However drying wetlands can have an impact on biodiversity. It is important to do an ecological survey if major plantings are planned.

Silvopasture systems can also be used to provide a biosecurity barrier between both herds and flocks on neighbouring farms (using wide boundary planting for example). The presence of a natural barrier can significantly reduce the transfer of diseases between flocks by stopping nose-to-nose contact.

Introducing trees into poultry systems improves poultry welfare and reduces stress for the animals. This, in turn, leads to increased production and higher-quality eggs[7]. It can also reduce the risk of poultry interacting with birds carrying avian influenza since the greatest risk comes from wild birds that congregate in more open landscapes.

Supplementing livestock diet

Trees generally contain higher levels of micronutrients than grasses. Tree fodder is a traditional livestock practice that has largely died out in the UK. There is renewed interest in the potential for using tree fodders particularly for addressing micronutrient deficiency and for their anti-parasitic properties associated with the secondary compounds (tannins) found in the leaves. Research in Holland showed that willow coppice introduced into a dairy system was preferentially browsed by the livestock. Whilst intake was generally low (0.6 and 0.4% of the required dry-matter intake for dry and lactating cows respectively), the intake of sodium (Na), zinc (Zn), manganese (Mn) and iron (Fe) was between 2-9% of the daily requirements[8]. These elements would normally be supplied through mineral supplements. There is also potential to use trees as a fodder source in drought events. Fodder can be harvested and stored for 24 months prior to use. Traditionally species such as elm, willow and ash were used.

Tree fodder is very common in tropical agroforestry systems, but the cost-benefit ratio is uncertain at present. As this is an area of active research there are, at present, potential risks that may outweigh the benefits offered by these systems. They are included here as interesting examples of areas where farmers are experimenting with silvopasture.

Broader environmental benefits

Trees provide a set of environmental benefits that can be realised off farm. Expanding the area of silvopasture, for example, will sequester significant amounts of carbon (C). These are higher in silvopastoral systems than they are in silvoarable.

Silvopasture can also provide biodiversity benefits. It is associated with higher soil biodiversity and provides semi-natural habitat and habitat networks for a range of birds, mammals and other fauna.

All of these interactions provide societal benefits which policy should support. These can benefit the farm through access to environmental grants through agri-environmental schemes.

Different types of silvopasture system

If we look at how trees are arranged within silvopastures then we can divide them into three broad categories detailed in Table 7 below.

Table 7: **Different types of silvopasture**

Silvopasture system			Examples
Trees within livestock pastures	A	Boundary tree systems	Shelterbelts, riverside planting and hedgerows
	B	Regularly spaced tree systems	Grazed orchards, row systems, clump systems
Livestock within woodland systems	C	Woodland grazing	Pannage systems, silvopoultry, parklands

Linear tree systems

Linear tree systems are used in silvopastoral systems primarily when some form of buffer is required – usually for protection of wind, soil and water quality. These systems will usually have timber, firewood, biomass and fruits (primarily from the shrub layer) as by-products. These systems are described in detail in Chapter 5.

Regularly spaced tree systems

These are systems where trees are introduced into pastures in regular patterns or in rows, normally with the intention of producing or maintaining a high-value wood product (timber or fruit). These systems are more likely to be successful in areas with better soils and lower exposure. In areas with high exposure clump systems can be used which provide better protection from the wind.

Other examples include apple orchards where sheep are introduced for part of the year. In orchards sheep reduce mowing costs, increase nitrogen cycling and reduce grazing pressure on other parts of the farm, potentially allowing an additional hay crop to be produced[9]. Tree fodder systems (where trees are deliberately integrated into rows to provide a supplementary feeding source) are also most efficient using these designs (to allow ease of management).

Woodland grazing

Wood pastures and parklands were once relatively common in the UK landscape and represent traditional agroforestry systems. Agricultural intensification in the UK has largely been at the cost of these woodland areas. Livestock traditionally played an important role in woodland management. Pannage systems, where pigs were herded in beech and oak woodlands, helped to produce viable tree crops whilst the pigs benefitted from interactions with the woodland (access to shelter and fodder). Farm woodlands still account for a significant amount of tree cover in the UK and remnant farm woodlands can and do provide a broad range of agroecological benefits to livestock.

New farm woodland is more likely to be established as part of a diversification strategy to deliver wood products. An example of this would be Forestry Commission Scotland's Sheep and Trees Forestry Grant Package[10] which aims to help farmers to establish viable timber production of farmland whilst providing silvopastoral benefits to livestock in the form of shelter.

Figure 13: **Cattle in Scottish farm woodland**

©Jo Smith, Organic Research Centre

Designing for livestock benefits

Reducing cold stress

The role that shelterbelt systems, hedgerows and other linear features play in reducing wind stress in livestock is well known. These benefits are still found where trees are integrated into pastures, particularly in row systems or clump systems that can incorporate a shrub or nursing tree layer. In addition to sheltering livestock the trees will also produce a microclimate that allows fodder to green up earlier in the year and to withstand drought conditions more easily.

Reducing heat stress

Mature trees with broad canopies provide the best shade. Typically these are perceived as competing with pasture so are often limited to field boundaries (as part of the hedgerow network) or as mature standards within fields. This can cause problems if animals only have limited shading options as it forces animals to congregate around them. This can create unhygienic conditions with poaching and other problems. Regularly spaced trees create more even shade, stop livestock clumping together and encourage more natural behaviour.

When considering using trees for reducing heat stress look at all areas within the farm where animals congregate for any length of time (after milking or at road crossing points) and make sure these areas have adequate shade provision. Access to effective shading is likely to be more important in fields with a southerly aspect.

Figure 14: **Shaded cattle**

©Jo Smith, Organic Research Centre

Encouraging natural behaviour

Livestock utilise well-designed silvopasture more evenly than open pasture and they function better as a group. This is partly because they can use the trees to hide behind and under, and partly because there is less competition, and therefore less stress, over important resources such as shelter and shade. Consequently, social interactions improve within groups of animals in silvopastoral systems. For example, in cattle 78% of all interactions are social licking compared to only 41% on pasture where there are few or no trees[11].

Figure 15: **Trees as scratching posts**

Tree trunks and low branches make good scratching posts to help maintain coat condition

©Lindsay Whistance, Organic Research Centre

The trunks of mature trees and low branches act as good scratching posts. For all livestock, daily grooming is important for keeping their coat in good condition. Rubbing against trees removes dead skin and hair. Access to rubbing posts is especially important for sheep with self-shedding fleeces. Although poultry maintain feather condition using their beaks, they preen more under tree canopies than when on open ground.

Silvopasture for ruminants

Shade and shelter are the primary benefit and as such all ruminants may benefit from all forms of silvopasture. Shelter is likely to be particularly important in upland farming systems or where farmers practise extensively grazed or 'New Zealand' style dairy systems, where the cows spend the majority of the year outdoors and are likely be out-wintered in all weather conditions.

Open regularly spaced silvopasture systems offer less shelter than shelterbelt systems designed for this function, though they give more even shade. These systems are more suited to lowland pastoral systems. It is possible to cultivate rape and stubble turnips for grazing by livestock in these systems as well as grass.

Farm woodlands can also provide shelter. Both sheep and cattle can benefit from access to woodlands although management varies because of their different browsing behaviour. Grazing woodlands poses several challenges to livestock management but also offers potential benefits to both ruminants and the woodlands. For a great overview on woodland grazing systems see Forestry Commission Scotland's Grazing Woodlands Toolkit[12].

Silvopasture for poultry

Silvopasture is an option for farmers interested in organic and free-range poultry systems where chickens have access to an outdoor run. Trees provide shelter to the chickens and they are more likely to engage in ranging behaviour, which has positive impacts on their welfare resulting in improvements to their health and production. See the case study on page 56.

Silvopasture for pigs

Pigs can benefit from access to woodland grazing. They particularly benefit from shade during the summer. Pigs are omnivorous and have access to a broad range of food types within forests (including roots, berries, nuts and plants). Their rooting behaviour can be used to reduce bracken cover, however their behaviour can be unpredictable and needs careful management. If stocking levels are judged carefully and kept low, their rooting action can be beneficial, reducing rank vegetation and encouraging seedling germination. Unmanaged they can have very negative impacts including complete loss of ground cover and damage to trees[13].

Using trees to reduce farm emissions

Trees can also play a practical role in mitigating the environmental impacts of both pig and poultry production on the farm. Studies have shown that tree shelterbelts downwind from farm structures can capture ammonia (which is heavily associated with both pig and poultry farming systems). Tree belts of 10 m width have been shown to reduce ammonia in emissions by around 53%[14].

**Chapter 3
Silvopasture**

CASE STUDY: **Trees mean better business**

David Brass, CEO of The Lakes Free Range Egg Company, is a recognised advocate of tree planting as an active part of farm management. David has found that for his business "there is no downside to planting trees".

As part of the McDonald's Sustainable Egg Supply Group, David worked closely with researcher Ashleigh Bright from FAI Farms Ltd to determine the effects that tree cover had on free-range flocks. Their report, published in the Veterinary Record in 2012, compared 33 flocks with tree cover to 33 without. It showed that chickens with tree cover produced eggs with better shell quality and reduced 'seconds' during collection and packing.

In November 2013, David secured a deal with Sainsbury's who strongly champion Woodland Eggs as a premium product.

Key facts

- It costs The Lakes Free Range Egg Company £2,000 per ha to plant, but payback is achieved in six months.

- Data proves that tree planting improves shell quality and can drive up the percentage of Grade A eggs by some 2%.

- Health and welfare benefits include reduced stress, lower levels of injurious feather pecking and improved conditions within sheds.

- Hen mortality can also be reduced, particularly if hens die trying to access houses in periods of panic.

- Trees draw surface water into the soil: this improves muddy conditions and prevents run-off of contaminants, such as phosphates, into water courses.

- Chicken sheds produce ammonia and tree planting can help intercept ammonia emissions.

- Planting at The Lakes Free Range Egg Company has had an immediate effect on wildlife and biodiversity, with barn owls and red squirrels now re-established on the farm.

©David Brass

This case study was compiled by the Woodland Trust, for more information see https://www.woodlandtrust.org.uk/publications/2015/01/trees-and-egg-production/

Maximising the value of the trees

Choosing the right tree

Establishing trees is expensive and time-consuming and potentially costly to reverse, so getting the right tree in the right place is vital. In upland farming systems, for example, trees need typically to cope with high exposure to wind and/or seasonal waterlogging. Hardier tree varieties (such as aspen, birch, rowan, sessile oak, blackthorn, Scot's pine and hawthorn) will do best, especially those of local provenance that are likely to be better adapted to the conditions. In these farming systems agroecological benefits are likely to be the primary driver for initial establishment (i.e. providing shelter or reducing foot rot) rather than the production of high-value tree crop. Poorly drained pasture together with overstocking can cause poaching and can increase the incidence of lameness. Adding trees can intercept run-off and reduce water collecting on pasture reducing poaching and associated issues[15].

Tree arrangements

Trees can be planted evenly at wide spacing with densities varying from 100–400 trees per ha depending on tree species used and the livestock system. For most species at these densities the tree canopy will not over shade the pasture for the first 12 years of establishment. Fast-growing species such as ash and alder may begin to shade early. Use shade-tolerant grass varieties or raise the crown to limit shading effects. Once the canopy starts to close selectively thin the tree crop to maintain the sward and allow growing room for the trees you want to keep. The thinning can be used for firewood or fencing timber. Early work with even-spaced silvopasture tended to use single tree species. However mixed tree species can provide different products through the thinning cycle.

In row systems, trees can be planted more closely together (in one or more rows) and different combinations of trees can be used to provide nursing benefits to the final timber crop. The rows themselves need to be wide enough to allow access. As with the evenly spaced tree systems the canopies will eventually begin to shade out the grass. This can be limited by planting the rows in a north–south orientation. These systems can produce multiple products (such as fruit, nuts and even browse). If the rows are being used for fodder or biomass then trees that are easily coppiced (such as alder or willow) can be used on much shorter rotations that will limit the shading effect.

Clumps have several potential advantages in terms of production and environmental impact over individual tree planting. The cost of tree protection is lower for clumps. Within clumps it is possible to select high-quality trees, as is done in conventional forestry, by progressive thinning to leave a small number of final crop trees in each clump. Furthermore, shading amongst trees within the clump may have silvicultural benefits of enhancing tree height growth and self-pruning and in exposed conditions the outer trees may shelter inner trees.

For environmental benefit, a micro-woodland habitat may be created in the clump with a richer wildlife value than that associated with single trees in fields. The shelter value of clumps can be increased by selecting trees that produce a dense, evergreen or early flushing cover around the edges.

In all cases mixed tree systems are likely to have better resilience to tree pests and diseases but may offer greater economic risk.

Figure 16: **Silvopasture clumps at Henfaes experimental farm at Bangor University**

©Jo Smith, Organic Research Centre

Light competition and grass growth

For the initial phases of tree growth there is usually very limited impact on forage growth but as trees mature they may need management (both pruning and thinning) to reduce the shading from the tree canopy to maintain sward quality. As the trees mature they reduce available light to the forage canopy but do bring other benefits. Trees shelter forage allowing it to green up sooner in spring. Similarly, the microclimate that trees provide protects the forage beneath them from heat stress. In this way silvopasture increases the resilience of the farming system.

Trees can also benefit the forage by accessing nutrients deeper in the soil and then returning this to the soil as leaf litter. If nitrogen fixing trees are part of the mixture of trees they may reduce or even negate the need for fertilizer.

Tree protection

Grazing animals damage tree stems, roots and ground vegetation and, as such, both cattle and sheep pose dangers to trees. Their natural behaviour is to trample (in the case of cattle) or browse and rub which means that establishment is impossible in most cases without protection and constant monitoring. If the rows have been left wide enough to allow vehicular access forage can be cut for hay or silage for the first few years until trees are large enough. This merely delays rather than removes the need for protection as once livestock have access to the pasture they will begin to damage the trees. There is significant cost to establishing any silvopastoral system. Sheep are the easiest to protect against, though still not cheap. The protections against cattle should be higher than those used for sheep and higher again against horses or wild ruminant such as deer. Protecting individual trees is more expensive than guarding rows or clumps of trees. Trees may also need protecting from poultry, and rabbits and voles can do damage to the lower trunk in any system.

Managing access through permanent or temporary fencing systems in existing woodland systems allows trees to establish by natural succession. This often means controlling the density of livestock in the woodland or removing them completely from sections of the wood. The grazing regime will vary with woodland types and livestock system.

Importance of management

Trees need as much management as pasture to flourish. In all cases weeding will be required around the base of each tree in the initial establishment phase (three to five years). Tree protection, where it is required, must be regularly checked, maintained and replaced if damaged. As trees grow they will need regular pruning during the winter months when the tree is less active to get the best-quality timber trees. This is a skilled activity for which training may be required. Trees with poor form should be selectively thinned. Tree management can produce a set of secondary products that can be used on farm including firewood, woodchip livestock bedding or Christmas trees.

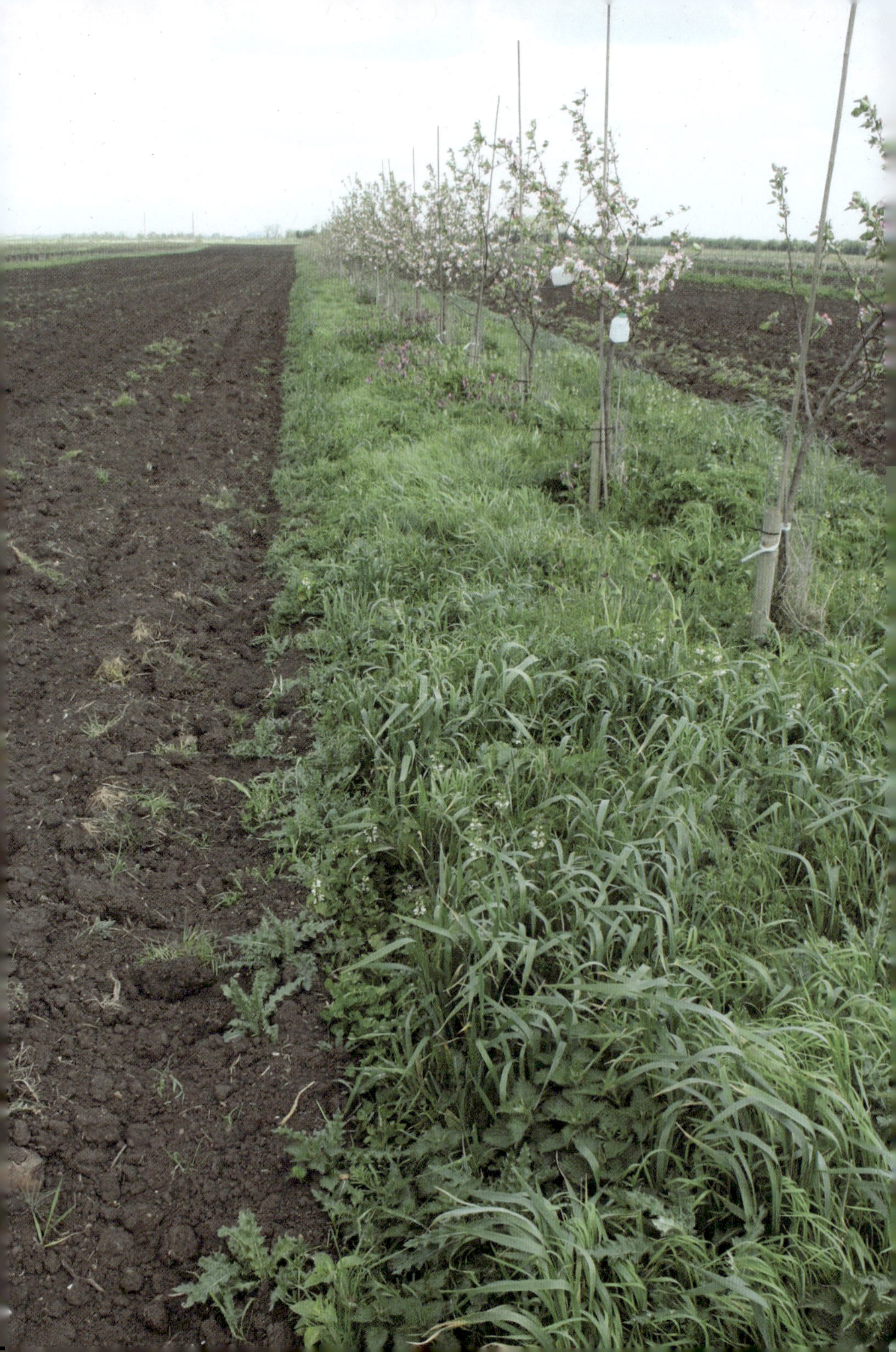

Chapter 4
Silvoarable
Dr Paul Burgess, Cranfield University

What is silvoarable?

Silvoarable agroforestry is the integration of trees with crops within the same field (see Chapter 1). The crops may be arable crops (e.g. wheat, barley and oilseed rape), horticultural crops and woody species such as short rotation coppice. Because of the need to allow continued mechanised management, the trees in silvoarable systems are usually planted in rows and the crops are grown in the intervening alleys. Hence another term that is also used for silvoarable systems is 'alley cropping'. This chapter examines typical objectives for silvoarable agroforestry and key design considerations. It then looks at options for maximising the value of i) the crop and ii) the tree products.

©Jo Smith, Organic Research Centre
▲Silvoarable at Home Farm, Nottingham

◄Apples and arable at Whitehall Farm, Cambridgeshire
©Stephen Briggs

Objectives and benefits of silvoarable forestry

In most cases, the starting point of a silvoarable system is an existing arable or horticultural system.

Designing an silvoarable system involves balancing a range of objectives. Is the principal objective to maximise crop production, enhance the environment, or maximise profitability from new tree products?

In a recent European survey, enhanced soil conservation and increased crop production were cited as the top two positive benefits of silvoarable agroforestry, with climate moderation ranked fifth[1]. Integrating trees in arable or horticultural systems can reduce wind speed, crop evapotranspiration and soil erosion. The loss and degradation of soils in the UK is an important concern; a recent study showed an annual cost of £1.2 billion for soil degradation in England and Wales. Almost half of the loss was related to the loss of soil organic matter, 40% due to compaction and 12% to soil erosion[2]. In organic systems, the inclusion of tree rows may also provide benefits for pest and disease control.

The European survey indicated that enhanced biodiversity and habitats were perceived as the third major benefit of silvoarable agroforestry. For example, measurements within a silvoarable system at the Leeds University farm increased the number of bank and field voles, wood mice and common shrews compared to an arable control area[3]. In turn these can be useful predators of insect pests and are themselves the prey of hawks and owls.

The fourth major benefit of silvoarable agroforestry in the survey was to diversify the sources of farm income from tree products. Examples of new products include timber, woodfuel, the sale of whole trees for amenity purposes, fruits such as apples, inflorescences such as elderflower and nuts such as walnuts.

Figure 17: **Elderflower planted in rows can be harvested for its flowers**

©Paul Burgess

Chapter 4
Silvoarable

What is the long-term plan for the system?

Is the aim to retain arable cropping over the length of a tree rotation (typically 25–60 years), or is the silvoarable system a way to ensure continued cropping and income during the establishment stages of the trees? Although many farms will have less structured forms of agroforestry systems Figure 18 shows one way of thinking about long-term planning.

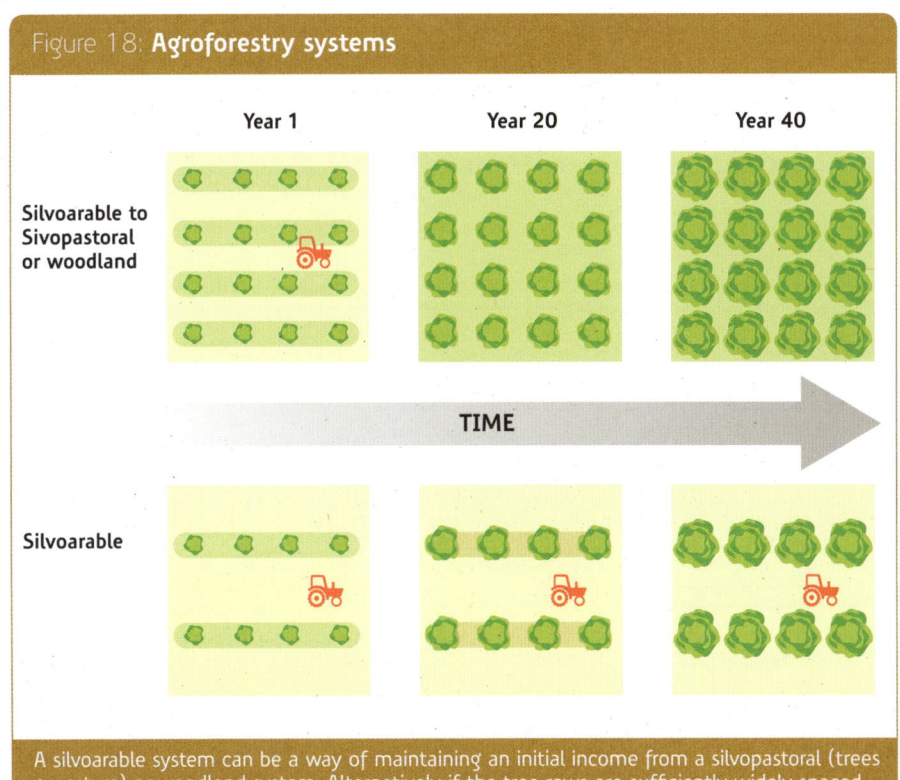

Figure 18: Agroforestry systems

	Year 1	Year 20	Year 40
Silvoarable to Sivopastoral or woodland			
Silvoarable			

TIME →

A silvoarable system can be a way of maintaining an initial income from a silvopastoral (trees + pasture) or woodland system. Alternatively if the tree rows are sufficiently widely spaced and/or if the trees are managed, the silvoarable system can be maintained indefinitely

Chapter 4
Silvoarable

Types of silvoarable systems

Silvoarable agroforestry for the establishment of a tree crop

The use of arable cropping to improve the cash flow of tree establishment was the basis of a poplar production system developed in Herefordshire and Suffolk by the match manufacturer Bryant and May in the UK in the 1960s and 1970s. They established poplar trees at a spacing that allowed profitable arable cropping in the initial years of tree growth. This system was part of the rationale for the UK Silvoarable Network of experiments with poplar at a 6.8 m x 10 m spacing that started in 1992. It included sites at Cranfield University at Silsoe (Figure 19A) in Bedfordshire, the Leeds University Farm near Tadcaster in Yorkshire and a site at the Royal Agricultural University in Cirencester in Gloucestershire. At the Silsoe site, arable cropping continued for the first 11 years until 2003. However the increased shading from the trees meant that the understorey was converted to pasture.

Figure 19: **Silvoarable agroforestry can be designed for A: the establishment of a tree crop or B: continued arable cropping**

A Silvoarable system at Silsoe, 10 years after tree planting
With narrow alley widths, the silvoarable system may evolve into a silvopastoral or woodland system

B Silvoarable system in France
With wide tree alleys as practised in France cropping may continue indefinitely

©Paul Burgess.   ©Arbre et paysage 32

Long-term silvoarable agroforestry for arable cropping

In contrast to the initial UK silvoarable network, managers of new silvoarable systems have generally opted for substantially greater alley widths that allow continued arable cropping as the trees grow. Here are few examples from Europe:

- Systems in Northern France have 28 to 110 trees per ha, with alley widths between 26 and 50 m (Figure 19B)[4].
- In Eastern Germany, there are experimental sites where 12 m wide hedges of short rotation coppice are planted amongst arable alleys that are 24, 48 and 96 m wide[5].
- In the Veneto region of Italy, silvoarable systems have been created by planting trees along open ditches spaced at an interval of 33 m and 90 m.

Design considerations

What are the appropriate tree density, orientation and spacing; both 'inter-row' spacing (between tree lines) and 'intra-row' spacing (between trees within a line)? The most critical decision is probably the inter-row spacing.

Inter-row spacing

To maintain arable cropping for the duration of the tree rotation, the distance between tree rows should allow continued profitable arable crop production. The alley width should be at least as wide as the widest piece of farm machinery such as a boom sprayer. To minimise 'double-working', the alley crop width should also be a multiple of the narrowest working width (e.g. a combine harvester or a seed drill).

For a long-term system, van Lerberghe (2017)[6] argues that the distance should be at least twice the eventual height of the trees. Hence with poplars reaching a height of say 15 m, the distance between rows should be at least 30–45 m.

The UK Silvoarable Network produced a simple model to predict the effect of alley width on crop yields[7]. Experimental results indicated that with trees spaced 10 m, and with side-pruning on the poplar trees to a height of 8 m in the first eight years, crop yields per cropped area could be maintained until year 10, but then declined sharply as tree pruning stopped (Figure 20). By contrast, crop yields per cropped areas were predicted to remain above 65% of the control with the 40 m alleys (Figure 20).

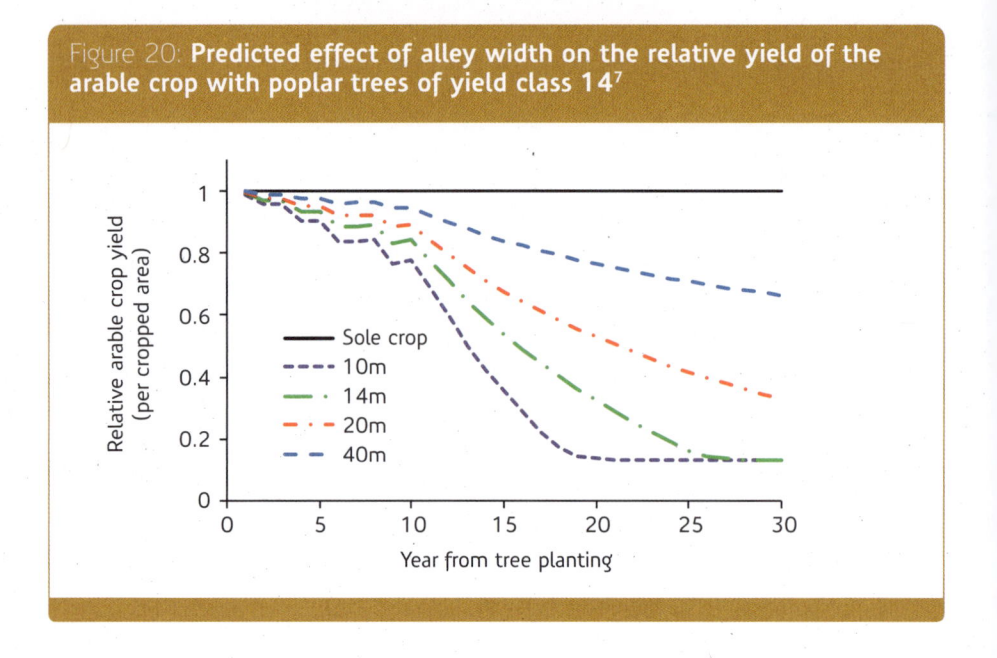

Figure 20: Predicted effect of alley width on the relative yield of the arable crop with poplar trees of yield class 14[7]

Intra-row spacing
In most silvoarable systems, trees are planted 4–10 m apart within the row. There is potential to remove (i.e. thin) the least productive trees early in the rotation. An example system practised in Italy is to plant alternating hybrid poplar and oak trees at an intra-row spacing of 7–10 m [8]. The aim is to harvest the poplars at 10 years, leaving the oaks to form a final timber crop.

Width of the non-cultivated strip
How wide should the uncultivated strip next to the trees be? Two metres appears to be the minimum width to avoid damage from machinery. More may be needed if machinery access is required at times when crops are still in the field, e.g. if harvesting apples when cereals are still in alley. Commercial systems with a single row of trees usually leave 2–4 m. However if there is more than one row of trees, then the width needs to be greater. For example, the woody vegetation rows were 11 m wide within an alley cropping system with short rotation coppice in Germany[5].

Turning area at the end of rows
Leave an area with no trees at the end of each row to allow access for machinery.

Orientation of tree rows

Researchers in France have used a 3-D agroforestry model to look at how tree row orientation affects light availability for alley crop at latitudes equivalent to the UK[9]. The model was run assuming walnut trees in alleys of either 17 or 35 m, growing to a height of 19 m with a crown radius of 8.5-10 m, and side pruning of branches to a height of 4 m. The modelled results showed a linear relationship between the reduction in solar radiation received by the crop and the diameter of the tree trunks. The results also showed that at latitudes found in the UK, a north-south orientation of the tree rows was better than a west-east orientation, in terms of reducing the variability of the solar radiation received by the alley crop and increasing light availability in the summer.

Tree row orientation can also affect wind speeds. If the aim is to reduce soil erosion by wind, plant rows perpendicular to the prevailing wind direction. For most of the UK, the prevailing wind comes from the south-west, although in places hill ranges can result in local differences.

A German study into the effects of providing wind shelter on arable crops found lower wheat yields within a distance of 1-3 m from the tree row[5]. However overall they reported a 16% yield increase in wheat yields in the alley relative to wheat in an open field, with the greatest increases observed 9-15 m away from the edge of the tree row. They related this higher yield to a 27% reduction in the potential evapotranspiration rate of the wheat in the alleys.

Combining the need to maximise light and the benefits of reducing wind speed, a north-south or a north-west-south-east orientation is likely to be most effective in Britain and Ireland. In practice, orientation of the trees also must consider field shape, orientation of open drainage ditches and slopes. On steeper ground, the soil conservation benefits may make it more appropriate to plant the trees along contour lines.

Risks to underground services and drains

Though often not as much of a problem as sometimes feared, consider underground services and drains. Poplar and willow roots can spread huge distances and cause problems in field drains. Consider planting rows of trees in line with the drain system if it exists in the field already. If a row of trees is planted directly above a drain the trees will only, over time, block that single drain and replace its function. If trees are planted across the direction of the underground drains then there is the potential to block every drain in the field. If an underground drainage scheme is linked by a main drain on the headland keep all trees at least 10 m away from that headland.

Avoiding electricity or telephone lines

Tree lines also need to avoid overhanging electricity or telephone lines (Figure 21).

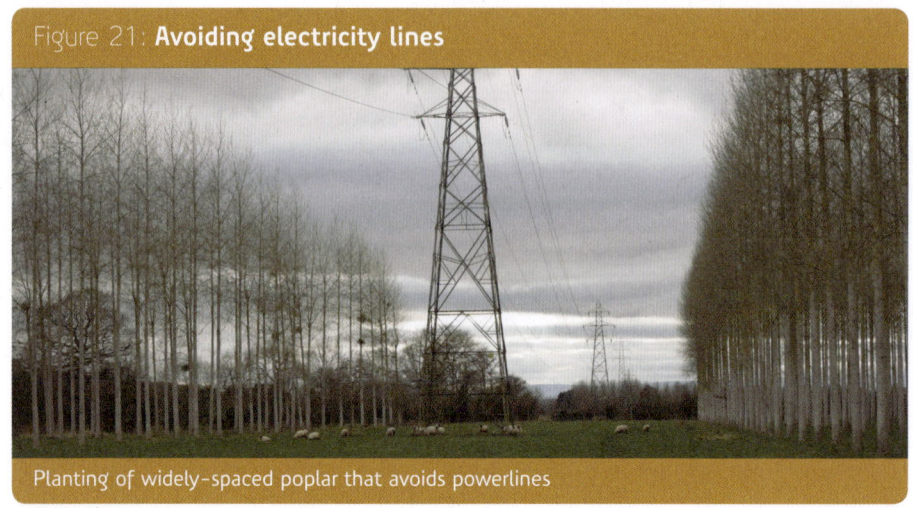

Figure 21: Avoiding electricity lines

Planting of widely-spaced poplar that avoids powerlines

©Paul Burgess

Novel designs

In some situations, a totally novel design may be appropriate. João Palma in Southern Portugal described the establishment of cork oak in a spiral silvoarable system, based on the width of the widest farm machinery (12 m) plus 1 m (Figure 22). The intra-row width was 2 m. The tractor driver initially commented "This seems a bit stupid, isn't it?". Four years after planting, both the farmer and the tractor driver are still pleased with design.

Figure 22: A spiral planting design used with cork oak

At Herdade da Torre do Lobo farm in Portugal

©João Palma, 2014

Chapter 4 — Silvoarable

CASE STUDY: Whitehall Farm – Planting to improve economic returns

Stephen and Lynn Briggs are tenant farmers at Whitehall Farm in Cambridgeshire. They have integrated trees into their wheat, barley, clover and vegetable-producing business, establishing the largest agroforestry system in the UK.

The system was implemented to reduce wind erosion affecting the fine grade one soils on the farm. It also enhances biodiversity, creates a mix of perennial and annual crops better able to meet the challenges of climate change, and diversifies their cropping.

Apple trees were planted in rows as windbreaks, but also to produce fruit and 24m alleys were left in between the tree rows for cereal production. A diverse range of pollen and nectar species and wildflowers has been established in the 3m wide tree understorey strip beneath the trees. This benefits pollinating insects and farmland birds.

The 52 hectare silvoarable agroforestry scheme cost an initial £65,000 to establish in 2009. In total 8% of the land is planted with trees and the remaining 92% is cropped under the existing cereal rotation. It took five years for the trees to mature into full production.

©WTML/Tim Scrivener

Key facts

- Trees can reduce wind erosion, while also enhancing biodiversity.
- Tree roots gather nutrients and water from deep in the soil, beneath the zone used by the arable crops.
- Adding value to commodities like cereals is difficult, whereas there is greater potential to increase the value of fruit through processing and direct sales.
- With the trees now seven years old, fruit yield per ha is similar to the surrounding arable crop, with gross margins typically c.£1000/ha.

This case study was compiled by the Woodland Trust, for more information see https://www.woodlandtrust.org.uk/publications/2017/06/whitehall-farm-planting-to-improve-economic-returns/

Maximising the arable benefits

As already discussed, planting trees on arable land can offer benefits to the understorey crops in terms of helping to conserve the soil and reducing wind speeds. However, the value of the arable crop in a silvoarable system can also be increased by choosing the correct type of crop, maximising light interception by the crop and minimising weed competition.

Choice of arable crop

A wide range of arable crops have been used in silvoarable systems. The crops tested in the UK Silvoarable Network included winter wheat, barley and beans, and spring wheat, barley and peas. It probably makes sense to avoid crops with a C4 photosynthetic pathway like maize which benefit from high light levels. Previous advice suggested avoiding potatoes, but they have been grown in an organic system at Wakelyns Agroforestry in Suffolk. Sugar beet had also been warned against due to the large machinery, but it has been successfully grown in Germany in 24 m alleys[5].

Figure 23: **Willow and barley silvoarable system**

Wakelyns Agroforestry, Suffolk UK

©Jo Smith, Organic Research Centre

Chapter 4
Silvoarable

Maximising light interception by the crop

The amount of light intercepted by the crop in a silvoarable system can be maximised by pruning the trees and choosing a crop where the light requirements are complementary to those of the tree.

Pruning: Pruning the trees can increase the light available to the arable crop. In some situations, such as fruit trees, the trees may be pruned when they reach a certain height. For timber trees, restrict pruning to encourage a main leading stem and the removal of side branches. The pruning of side branches increases both the volume of knot-free timber and the light available to the understorey crop. The predicted benefit of side-pruning branches of the poplars grown in the UK Silvoarable Network site is shown in Figure 24. Pruning the side branches on six occasions to have a branch-free trunk to a height of 8 m reduced the width of canopy development and extended the period where crop yields were close to that in the control plots by about seven years.

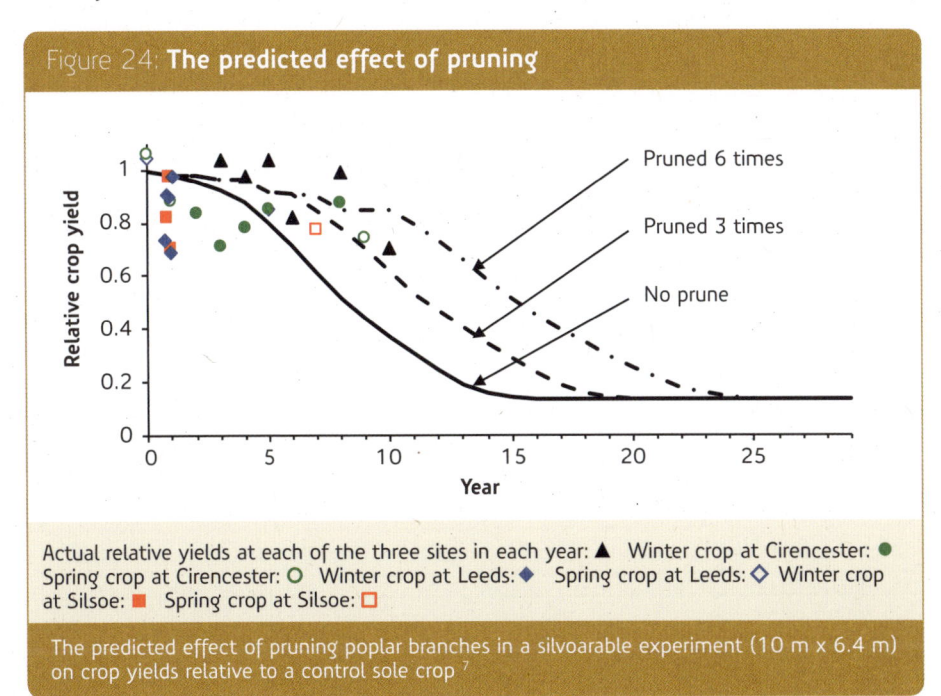

Figure 24: The predicted effect of pruning

Actual relative yields at each of the three sites in each year: ▲ Winter crop at Cirencester: ● Spring crop at Cirencester: ○ Winter crop at Leeds: ◆ Spring crop at Leeds: ◇ Winter crop at Silsoe: ■ Spring crop at Silsoe: □

The predicted effect of pruning poplar branches in a silvoarable experiment (10 m x 6.4 m) on crop yields relative to a control sole crop [7]

Choosing complementary crops: Some tree species such as poplar only achieve full light interception late in the growing season; for example the light interception of poplar at the UK Silvoarable Network site at Silsoe was only achieved in late June (Figure 25). By contrast an autumn-sown wheat crop can intercept significant amounts of light in April, May and early June. Hence an autumn-sown crop will have a more complementary (i.e. less competitive) light-capture pattern than a spring-planted crop, when planted with most deciduous tree species.

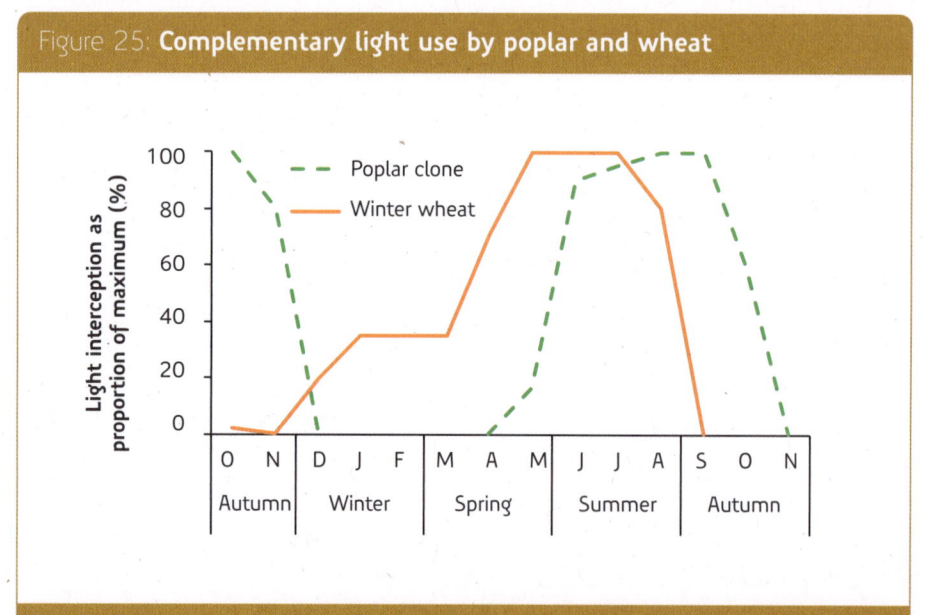

Figure 25: **Complementary light use by poplar and wheat**

The light interception of the poplar hybrid Gibecq and a winter wheat crop is complementary. The winter wheat crop can use light when the poplars are dormant from December to May, and the poplar can use the light when the crop is harvested (after Incoll and Newman, 2000[10], and Pasturel 2004[11]).

Minimising weed and pest competition from the tree row

The risk of increasing weeds is a concern when planting a new agroforestry system. One study from Northern France showed that tree rows had no obvious negative effect on the distribution of weeds on organic farms. Tree rows have, however, been shown to increase weeds on farms and experiments using agrochemicals[12].

On the UK Silvoarable Network sites, weeds in the tree rows were initially controlled using a black plastic mulch. However, as the plastic disintegrated, the tree row became colonised by arable weeds such as barren brome, blackgrass and common couch[7]. Common couch also developed in the tree rows during the first two years after tree planting on the organic vegetable silvoarable system managed by Iain Tolhurst in Berkshire[13]. Weed measurements at the Leeds and Cirencester UK Silvoarable Network sites showed a higher presence of weed species and a greater cover of weed species in the cropped alley next to the tree row than found in the open field[7].

Various methods of controlling weeds in the tree alley have been tried including the use of buried black plastic sheeting, non-selective herbicides like glyphosate[7] and the use of organic mulches. Black plastic mulches were initially effective, but the eventual removal of the plastic was labour intensive. On the UK Silvoarable Network sites, it was possible to establish a grass mix including cocksfoot and red fescue following removal of the black plastic[7] and this reduced the number of weed species within the understorey. However even then couch grass and blackgrass remained problems on the clay soil at the Cranfield site at Silsoe. It has been suggested that planting a wildflower mix in the tree row may be a more productive means of controlling the extent of aggressive weeds.

The effect of the tree row on pests and diseases is less clear. Vegetated tree rows can provide a refuge for spiders and ground carabid beetles, which may offer some benefits for pest control within the arable crop[7]. However, the tree row can also create problems; for example, in an experiment at Leeds University, Griffith et al. (1998)[14] associated lower crop yields close to the tree row with slug damage.

Chapter 4
Silvoarable

Growing an additional crop in the tree row

One possible way to further increase revenue from a silvoarable system is to grow a commercial understorey crop in the tree row. Adolfo Rosati in Italy has promoted planting wild asparagus in the tree rows of an olive system. In the UK, Iain Tolhurst has trialled planting rhubarb and wild flowers, such as daffodils, within the tree-row of an organic system[13].

Maximising the value of the trees
Tree selection

Chapters 2 and 6 contain information to help choose the right tree for your farm, in this chapter we will look at some that are more suited to silvoarable systems. The initial UK Silvoarable Network with sites at Cranfield, Leeds and Cirencester only focused on the use of poplar hybrids. Although poplars grow very quickly, poplar wood is softer than most broadleaf species and it is not easy to find profitable markets for the timber. One of the most successful examples of maximising the value of trees within an agroforestry system has been the production of ash trees (albeit in a silvopastoral system) at Loughgall in Northern Ireland. During the first thinning of the trees, the ash was sold for hurley sticks. Unfortunately, since then, ash dieback has reached the UK and forest authorities no longer recommend planting ash.

In France, walnut has been successfully used in a silvoarable system such as the system at Les Eduts practised by Monsieur Jollet. A financial analysis of silvoarable systems across Europe in the early 1990s also highlighted walnut with arable crops in France as one of the most profitable systems[15].

Because of the absence of one particularly lucrative tree species, the most common procedure in alley cropping systems has been to plant either apple trees or a mixture of tree species. This minimises the risk of the complete loss of all trees due to pests and diseases.

Stephen Briggs at Whitehall Farm near Peterborough established nine commercial and four traditional apple varieties. Farmers in Northern France have been planting six to 12 species per field including common walnut

(*Juglans regia*), Norway maple (*Acer platanoides*), wild cherry (*Prunus avium*), wild service tree (*Sorbus torminalis*), service tree (*Sorbus domestica*), apple and pear species, sycamore (*Acer pseudoplatanus*) and black locust (*Robinia pseudoacacia*). See the case study on page 69.

Iain Tolhurst in the UK, in an organic silvoarable system planted 18 varieties of apple, field maple (*Acer campestre*), whitebeam (*Sorbus aria*), Italian alder (*Alnus cordata*), oak (*Quercus robur*), black birch (*Betula lenta*), hornbeam (*Carpinus betulus*) and wild cherry (*Prunus avium*).

At Wakelyns in Suffolk, Martin Wolfe established rows of willow short rotation coppice. This produces a biomass energy crop within four years of planting, and once established it can be harvested every two to three years, or five years with hazel. It should be noted that to minimise the effect of insect and disease damage, growers are recommended to grow a mixture of willow varieties.

Maximising tree growth by minimising water competition

It is often assumed that because trees are tall they can compete well with arable and grass crops. Young trees can compete well for light, but the root density of young trees is typically an order of magnitude smaller than grass or wheat and therefore young trees are susceptible to water and nutrient competition from any vegetation in the tree-row and even the arable crop in the alley. For example, results from the UK Silvoarable Network site shows that after seven years of growth, trees growing next to an arable crop were about 10% shorter than those surrounded by land that was kept continuously fallow, and the tree diameter about 20% smaller[16]. This was attributed to competition for water.

For this reason, early tree survival and growth is closely linked to minimising weed competition for water and nutrients. In the UK Silvoarable Network, weed competition against the trees was achieved by using black plastic mulch, but this was later difficult to remove. Other methods for minimising water competition include the use of a broad-spectrum herbicide and woodchip mulches.

Tree protection

Trees in silvoarable agroforestry need less protection from livestock than those in silvopastoral systems. However the trees still need protection from rodents and deer in most situations, hence tree guards are necessary. For example Iain Tolhurst, who has established an organic silvoarable system, found it was necessary to use large wire meshes around apple trees to reduce damage caused by deer[13]. Roosting birds can also damage the leading shoot of trees, hence Stephen Briggs in Peterborough had to place canes alongside the trees to prevent roosting. Stephen also found that when it snowed his rabbit/hare guards were too short as the animals would stand on the compacted snow to reach the trunks. He had to go around and add an extra layer of mesh to each tree to protect them.

Minimising herbicide damage to trees

There is a risk that herbicides applied to the arable crop can affect the trees when they are in leaf. For example, a non-selective herbicide like glyphosate applied to the crop area may also be taken up by the lower leaves of a tree.

Tree pruning and removal of epicormics

"If only the trees were correctly pruned," foresters frequently say. The highest timber prices are typically secured for wide knot-free trunks. Side-pruning the lowest stems means that subsequent growth can be knot-free. However an important second reason for side-pruning is to ease machinery movement within the alley. Unlike in a dense forest stand with low light conditions, the lower branches of a silvoarable tree will not self-prune. Hence the need for and costs of pruning per tree are typically higher in a silvoarable than a conventional woodland. However it should be noted that the number of trees per hectare is typically lower. When tree pruning, remove the prunings from the field. This is to prevent remaining branches causing damage to the farm machinery used in the alleys.

Chapter 5
Hedges, windbreaks and riparian buffers

Dr Jo Smith and Sally Westaway, Organic Research Centre

Agroforestry systems such as hedgerows, windbreaks and riparian buffers are widespread landscape features in the UK, providing a range of benefits for the farming system as well as the wider environment. In addition to discussing the main considerations for planning, planting and management, this chapter also presents options for managing these features as a productive part of the farming enterprise.

Although historically hedgerows, windbreaks and riparian buffers may have been planted for different reasons, they provide similar services to the farm and the environment, depending on their location and management. Boundary hedgerows are usually established to mark property or field boundaries, to improve the husbandry of livestock and to prevent damage to arable crops. In the past they were also managed as a source of food, materials and firewood. Windbreaks, or shelterbelts, are strategically planted strips of trees that aim to reduce wind speeds in the protected area. The main function of riparian buffers is to protect water courses by capturing sediment and nutrients from adjacent fields, buffering water courses from pesticide spray drift as well as providing shade, and buffering water temperatures to the benefit of river wildlife.

When positioned correctly all three features can reduce wind speeds in an area up to 30 times their height[1]. This reduction can have multiple benefits including increased crop growth rates and quality, protection from windblown soil, moisture management and soil protection. Higher air and soil temperatures in the lee of a windbreak or hedge can extend the crop growing season, with earlier germination and more growth at the start of the season. Fruit and vegetable crops are particularly sensitive to wind stress and suffer reduced yields and poorer quality at lower wind speeds than combinable crops. For livestock, reduced wind speeds and chill factors can increase live weight gain and milk production, reduce feed costs and young stock mortality. During the summer, by providing shade, trees can reduce the energy needed for regulating body temperatures, and so also result in higher feed conversion and weight gain.

These landscape features can also aid livestock management, as physical barriers between fields or farms. They can increase biosecurity by reducing contact between herds or flocks and, where livestock is excluded from wet areas of pasture, liver fluke and lameness may be reduced. By providing shelter in handling areas, working conditions can be improved for animals and humans alike.

Windbreaks, riparian buffers and hedges can reduce soil erosion from water, by reducing soil compaction and increasing infiltration rates and reducing overland flow of water; and from wind, by slowing wind speeds and reducing the energy available to dry and move soil particles. In addition to benefitting the farm (by retaining soil in fields), this also benefits the wider environment, by reducing sediment in streams and by reducing pollution run-off and flooding by increasing soil water storage ability.

In some landscapes, these features form the most widespread semi-natural habitat. They can therefore play an important role in supporting biodiversity on the farm, as well as linking up patches of woodland or other habitats to allow wildlife to move through the landscape. By providing shelter, food and nesting resources for wildlife, important services such as pollination and pest control may be improved in the adjacent field and beyond.

Figure 26: **Biodiversity in a Berkshire hedgerow**

©Jo Smith, Organic Research Centre

Chapter 5
Hedges, windbreaks and riparian buffers

Figure 27: Layed hedge

Layed hedge at Elm Farm, Organic Research Centre, Berks

© Organic Research Centre

Hedges, windbreaks and riparian buffers all need some management to persist and to continue to provide their function in the landscape, as well as to manage the interactions between the trees and adjacent fields. In addition to regular cutting to manage encroachment into neighbouring fields or roads, hedges need periodic rejuvenation actions – either by coppicing or hedgelaying – to encourage multiple stems, to maintain the hedge structure and function. Windbreaks and riparian buffers also need on-going management, such as thinning, coppicing or pollarding, selective felling and restocking in order to maintain their protective function. Some funding to cover the costs of such management may be available via governmental support (in recognition of their importance for the environment), but it may also be possible to manage these features as a productive part of the farming system. This can help to offset the costs of management, while also supporting the cultural, biodiversity and environmental values of these landscape features.

Chapter 5
Hedges, windbreaks and riparian buffers

Site selection, design and establishment
General guidelines

Any new planting should start with a review of existing woody resources and features on the farm, where necessary bringing these features into active management, e.g. gapping up, rejuvenation through coppicing or pollarding (where appropriate), and thinning and replanting where necessary.

Being realistic about the management implications of planting new trees is important, both during the establishment phase and in the longer term. While there are many benefits, trees and hedges can also compete with crops and grass for light, nutrients and water, as well as for farmers' time. Management to minimise this competition, and to ensure the long-term survival and functioning of these systems, should be carefully thought through at the design stage.

Planting linear features across characteristically open landscapes, wetlands, marsh, or unimproved grassland should be carefully considered as there may potentially be a negative effect on the landscape's character and wildlife. Some species, such as the lapwing, prefer wide open spaces and they may become vulnerable to predators such as crows and foxes attracted by hedgerows.

Figure 28: **Eastbrook Farm, new shelterbelt planting**

©Ben Raskin

Hedges

Site selection

The siting of new hedgerows will depend on the objectives (e.g. marking a new boundary, for biodiversity, providing shelter, screening a footpath or building, or providing fuel for the farmhouse). Where possible, plant on existing field boundaries or join up gaps in the hedge network or wildlife habitats. If practical, reinstate historic field boundaries; old maps of the farm will show where these are. Look in local archive offices if you don't have them yourself. There was a reason why they were sited there in the first place and there may be an opportunity to enhance the historic landscape character through new plantings. If planting new hedges for production, access for management and harvesting is an important consideration, e.g. planting a new woodfuel hedge alongside a farm track would allow regular coppicing independently of soil conditions. An excellent resource for anything to do with hedges can be found on the Hedgelink website www.hedgelink.org.uk

Design

For a stock proof hedge, aim for four to six plants per metre in staggered double rows, usually 40 cm apart. Traditional hedgerow mixtures typically consisting of native species such as blackthorn (*Prunus spinosa*), hawthorn (*Crateagus monogyna*) and hazel (*Corylus avellana*) will provide great resources for wildlife, as well as providing a stock-proof barrier and a dense physical boundary feature if managed properly. Planting species that are commonly found in local hedgerows is a good guideline, as they are likely to do better in the local climate and soils and fit the landscape character of the area. Mixed species hedgerows are valuable for wildlife, and typically consist of around 60% of one dominant species with a mixture of other species in varying percentages. In some regions, single-species hedges are characteristic of the local landscape.

If the aim is to also provide a product such as woodfuel or woodchip livestock bedding, however, thorny species such as blackthorn or hawthorn should be avoided. Faster-growing species, such as willow, hazel and even sycamore, could be planted, although it is important to recognise that these species will have a bigger impact on adjacent fields, especially if allowed to grow tall for maximum biomass production. Beneficial impacts (e.g. shelter from wind, or income/cost savings from the hedge product) may balance negative impacts on crop and grass yields in adjacent fields. Planting mixed-species productive hedges is better for wildlife (and may reduce potential pest and disease problems in the trees) but is only recommended if the species have similar growth rates.

Hedgerow trees are important landscape features. They provide shelter, food and nesting sites and make a valuable contribution to the landscape. Species choice could be influenced by the potential for timber (oak, wild cherry or beech) or other products (e.g. fruit trees) or by the species present in the surrounding landscape.

When designing a new hedgerow, windbreak or riparian buffer, be aware of any regulations regarding planting near roads, electricity lines, water courses or protected habitats.

Establishment

Bare-rooted-whips 40–60 cm are most commonly used for hedge planting as they are the most successful at establishing. Hedgerow trees are best planted at 6–10 m spacing at the same time as the hedgerow; taller whips, 1–1.5 m tall, usually establish better than larger trees if planting into a new hedge, but larger trees might be better if planting into an established hedge. Tree tags can be used to identify the hedgerow trees, to avoid them being cut with the hedgerow. Control weeds for the first few years after planting using mulches (e.g. woodchip, fabric or polythene mats or sheets) or herbicide and gap up to replace any dead plants. If gapping up an existing hedge, coppice or cut back the adjacent plants. Use shade-tolerant species (e.g. holly) if planting under a hedgerow tree. Protect the newly planted trees from browsing using tree guards and stock fencing where livestock is present. Stakes should also be used to hold the guards in place and to support the young whips. Guards should be removed once the hedge plants are well established to allow side branching at the base and prevent gaps forming.

To encourage bushy growth, the tips of the new plants can be cut back by one third, and the hedge trimmed lightly every second or third year, allowing the hedge to increase in size each time. If the hedge is to be laid or coppiced, however, just trim the plants up the sides until the leaders have reached a suitable height.

Hedgerows – The management cycle and planning

Many hedgerows are in decline, through under-management, mismanagement or removal. Of those that are still actively managed, the majority are repeatedly flailed at the same height, eventually creating gaps and poor hedge condition. Those left unmanaged ultimately develop into lines of trees. Both over- and under-management are detrimental to the structure of the hedgerow.

The Hedgerow Management Cycle[2] is a useful starting point for assessing the potential of farm hedgerows, deciding on appropriate management methods and developing a management plan. It is a 10-point scale based on the physical characteristics of the hedge that goes from 1 (an over-trimmed short hedge with many gaps) through to 10 (a line of mature trees). For management, hedges are best assessed in winter when the leaves are off and the structure can be seen; however it is easier to identify hedge species in summer.

Every hedge is unique, and the most appropriate management will depend on the hedge itself, its role within the local landscape and the farm, and the priorities of the landowner/farmer. The development of a hedge management plan for the entire hedge network on the farm is recommended to coordinate activities at a farm level.

A 'healthy' hedge is thick and bushy, with many interwoven branches that provide excellent shelter for wildlife (point 5 on the scale). To maintain the hedge in this condition for as long as possible, trim the hedge on a two- or three-year rotation, raising the cutting height incrementally at each cut. Eventually, however, the hedge will become gappy at its base and need rejuvenation through coppicing or laying.

There are opportunities to adapt traditional productive hedgerow management techniques for modern farming systems, so farmers can both diversify income streams and increase system sustainability. Hedgerow products could include bioenergy, soil-improving mulches (e.g. ramial woodchip or woodchip compost), fruit and nuts, livestock fodder and timber.

Hedgerows for woodfuel

Hedges can provide a local sustainable source of woodfuel. In some areas of France, hedgerows are still an important fuel source, producing 4.4 million m³ of fuel per year and accounting for 11% of the total annual firewood used by households[3]. Coppicing or hedgelaying are both rejuvenation methods that can produce woodfuel as a by-product, either as logs or chipped for use in biomass boilers. By managing existing landscape features such as hedgerows for bioenergy, farmers might not need to choose between producing food or energy from their land.

Hedgelaying will produce some material that can be used for fuel but a lot of the woody material will be retained in the hedge, producing over time a hedge which is thick, dense and excellent for wildlife. With the right management a hedge provides a stock-proof boundary without the need for an additional fence. Coppicing produces more material for use in an on-farm biomass boiler, or logged, either for use on-farm or as a secondary income stream for the farm. The hedge that regrows after coppicing has a different structure to a layed hedge; it is less bushy with more straight stems.

Figure 29: Hardwood cordwood before cutting and splitting for firewood

©Soil Association

# Introducing a coppice cycle to hedges

It is best to start by mapping the hedges and identifying any unsuitable for coppicing (e.g. historical or wildlife value or other functions such as visual screening). The remaining hedges can be assessed in terms of size and species composition and a coppice rotation planned. As a general rule, no more than 50% of hedges on a farm should be managed as part of a coppice rotation and no more than 5% of hedges should be coppiced in any one year. It is also important to consider landscape connectivity by maintaining or improving linkages between habitats such as woodlands and ponds.

Recent trials in the UK[4] assessed the feasibility of mechanising the process of coppicing hedges and chipping the resultant material. The trials concluded that hedges can be managed to produce woodfuel of a quality that meets industry standards.

Many different machines and combinations of machines can be used for harvesting hedgerows, ranging in scale from manual chainsaws, to mid-scale tractor-mounted circular saws, or excavators with tree shears, through to larger-scale machinery used in forestry such as felling heads and grapples. The choice of machinery option will largely depend on the type and length of hedge being coppiced.

If there is only a short section of hedgerow to harvest (less than 100 m) it may be more economical to use smaller-scale options such as chainsaws or small-scale tree shears, and a manually fed disc chipper. If using larger-scale machinery options and a crane-fed drum chipper, make sure there is enough hedge length and material (around 250 m) to keep hired machines busy for a full day. More information can be found in the Guide to Harvesting Woodfuel From Hedges[5].

Speed of regrowth following coppicing depends on species and the age and condition of the hedgerow tree at coppicing. Fastest regrowth will be for species that respond well to coppicing e.g. willow, hazel, alder, ash. Regrowth will be slower in exposed situations or on poorer soils, and protection from browsing animals should also be considered. Mixed-species hedges, depending on the species, may get variable rates of regrowth which could cause management issues in the future and may be better suited to manual coppicing, e.g. with a chainsaw rather than with larger-scale machinery.

Coppicing in the winter is best, both horticulturally and in terms of woodfuel production, when there are no leaves or green material present in the woodchip. The coppice stool responds well to coppicing with good regrowth, and there is no risk of disturbing breeding birds or animals. However, there may be logistical problems associated with coppicing in the winter. Firstly, most hedgerow and woodland contractors are very busy, and the availability of specialist machinery can be very limited. Secondly, ground conditions can deteriorate rapidly after mid-October depending on the location and soil type of the farm, with the potential for rutting and soil compaction.

Where coppicing can be done from the road or trackside, it does remove the need to track across fields but there is often not enough space for both the feller/machine and the hedge material, and road-closure applications may need to be made to the local authority.

The right boiler needs to be able to cope with the variable nature of hedgerow woodchip. The limited volumes and bulky nature of hedge biomass means that management of hedges for woodfuel is more suited to smaller decentralised short-chain energy-production systems. Farmers are well placed to establish local firewood or woodchip enterprises. Being locally based minimises transport costs and therefore can reduce firewood and woodchip prices and provide rural employment.

Include the cost saving of reduced flailing when doing your economic assessment of hedge coppicing as well as the potential for government support via environmental stewardship payments.

Managing hedgerows for other products

Woodchip from hedgerows has many potential uses on a farm from livestock bedding to use as a soil improver, as compost, or as mulch for weed control. When chipped for compost, as a soil improver or for mulch, the quality of the chip is less important than when used for fuel.

Woodchip can be turned into compost in as little as three months to one year depending on the frequency of turning, and the inclusion of some green waste. There is also evidence to suggest that the application of a thin layer of uncomposted (ramial) woodchip at an appropriate phase in a crop rotation can increase soil organic matter, water-holding capacity and nutrient levels of soils; however, research on this subject is currently limited. Young branches are nutritionally the richest parts of trees, as they are exposed to the most light and are the most actively growing. As such, material harvested and chipped from smaller tree branches or hedges provides ideal material for the production of ramial woodchip.

Woodchip could also provide an alternative bedding material to straw, particularly where straw is in short supply, and may offer many animal health and welfare benefits, with limited bacterial growth and less dust than straw. Chip needs to be dry (not above 25% moisture content) and if produced on-farm should be dried for six to 12 months before use. Species with thorns should be avoided (e.g. blackthorn, hawthorn), but most other seasoned hard and soft woods will work equally well as bedding, although larch should be avoided due to its tendency to splinter[6]. Using larger chips allows liquid to pass through to lower layers, leaving the upper layers relatively dry and friable. The AHDB recommends a shallow 10 cm depth with a fresh top-up layer applied as required (typically every seven to 10 days if animals are on a dry diet, more frequently if fed a silage-based ration)[6].

The used material can be composted (heaped and turned every 4 to 6 weeks) and the resulting material sieved to separate coarse woodchips to be re-used as bedding, from compost which can be spread on land or composted further.

There are many other potential products that can come from hedges, such as fruit (from fruit trees planted in the hedge through to blackberries and the marvellous sloes for sloe gin!), binders and stakes for hedgelaying, fence posts and timber. Ideally, such new products would be used on farm, or complement what is already produced (e.g. new lines of fruit or vegetables in a horticultural enterprise). Alternatively, new markets may need to be sought or interest generated for the new crop within existing markets; some creativity may be needed (e.g. direct selling or adding value to produce by making jam). The labour requirements for on-going maintenance and harvesting need to be considered. It's time to look with fresh eyes at your boundary hedges and get creative!

Figure 30: **Edible hedge with almonds**

©Sally Morgan

Windbreaks

Site selection

Where to plant windbreaks will be largely determined by environmental and geographical factors. The wind direction(s), topography and farming practices should all be considered at the planning stage. Identify which areas need protection, which areas are particularly prone to soil erosion by wind or water, and the prevailing wind direction (or most damaging wind direction if there is more than one prevailing direction). Windbreaks sited at right angles to the prevailing (or most damaging) wind give maximum protection. However, also look at linking up existing woody patches on the farm and opportunities to improve livestock management. Plant trees as barriers to prevent disease spread between fields and farms, or to shelter handling areas.

Design

The effectiveness of a windbreak is influenced by its height, length, orientation, continuity, width, cross-sectional shape and permeability. To maximise the area sheltered, the windbreak should be as tall as possible, although this may conflict with other considerations such as shading impacts on crops. Permeability is particularly important – dense barriers force the wind upwards, creating high levels of turbulence where the wind returns to the ground. Approximately 40% permeability is the most effective. This can be achieved through manipulating tree densities, number of tree rows and species. The optimal width of the shelterbelt will vary, depending on the shelter required and species used, and taking into account the need for permeability; the wider the windbreak, the less permeable it is. To encourage the wind through the trees rather than deflecting over the top (and causing turbulence) the best design is one or more lines of trees in the middle, with shrubs either side which can be kept trimmed back. Wind coming round the side of a windbreak encroaches on the area sheltered, so it is recommended that the length should be around 10-12 times the height. Also consider the length of windbreak relative to livestock densities to reduce crowding and poaching. Avoid gaps in the windbreak which can create wind tunnels; if needed for access, planting small islands of trees upwind of the gap can mitigate this.

To achieve shelter quickly, fast-growing species such as poplar, alder or birch may be planted initially; these will provide shelter so that other longer-living species such as oak can get established. A species mixture promotes an irregular canopy height which helps reduces wind eddies. As the main trees mature and thin out, it is important to maintain shelter at lower levels by planting shade-tolerant shrubs, or by coppicing understorey species such as willow and hazel. Trees planted for timber should be planted in the centre of the windbreak to avoid side branching.

Establishment

Planting densities will depend on the choice of species and windbreak design. Standard densities tend to be at least 2,500 trees per ha (i.e. 2 m apart) with fence lines at least 1 m from the edge of the windbreak.

Windbreaks – Management for protection and production

Once established it is tempting to think that a windbreak will get on with its job without the need for further care. But as the trees age, the windbreaks may first become too dense (i.e. permeability decreases) and then become too open as growth slows and trees die. To keep the porosity optimum, and provide on-going shelter, management is required. Thin the trees after 15–20 years. This gives trees more space to grow, and if the aim is to also produce a timber crop, this will improve yield in the long term, while producing a crop of woodfuel or fencing stakes in the short term. Thinning also reduces the risk of wind blow and gives slower-growing species the opportunity to get established. Coppicing is also an option, and the regrowth will provide more shelter underneath the canopy.

When the trees reach maturity, you can clear fell the windbreak and replant, but this compromises the provision of shelter as the new trees establish. It should be possible to maintain shelter while removing the mature trees. Options include cutting and replanting first the leeward half of the windbreak, and once that has established, do the same with the windward half. Alternatively, if land is available, a new windbreak can be planted adjacent to the existing one; once established the old one can be removed. If there is enough light reaching the floor of the windbreak, it may be possible to plant with shade-tolerant species such as holly or hazel, or if the windbreak is large, selective felling of groups of trees can open up patches of light in the canopy which can then be replanted. Finally, depending on the level of browsing by wildlife or livestock, natural regeneration may occur, allowing understorey trees to establish and replace older trees over time.

Riparian buffers

Site selection

Environmental factors and farm geography are important for riparian plantings too, but you should also consider the main function for the buffer. If the aim is to reduce run-off from fields, buffer strips will be sited at the interface of the field and water course. However, if the focus of riparian planting or regeneration is to reduce in-stream temperatures, siting buffers around headwaters and small water courses will be more effective as the water is more responsive to shading, while riparian trees further downstream may create cooler patches for fish to retreat to. More information on planting riparian buffers to control water temperatures can be found in the guide Keeping Rivers Cool: A Guidance Manual – Creating riparian shade for climate change adaptation[7]. There may be constraints on where you site riparian buffers, depending on the status of the water course and adjacent area and it is advisable to discuss planting plans with the local office of the Environment Agency, Natural Resources Wales or the Scottish Environment Protection Agency.

Design

Design buffers to hold water for as long as possible so they have the maximum impact on run-off and pollution; width of the buffer, slope, amount of vegetation and leaf litter, and soil type are all important. Water flows too fast on slopes more than seven degrees for riparian buffers to be effective. Buffers of between 5–30 m width have been found to be at least 50% effective at protecting the various stream functions. Imitate native riparian woodland with an open canopy of mixed species and with varied ages. There should be enough light to support a cover of herbaceous ground flora and vegetation along the water margin, and around 50% of the stream surface should be open to sunlight with dappled shade in the remainder. One option is to use natural regeneration to create the riparian buffer, but if planting from new, species with native light-foliaged species should be considered, such as birch (downy and silver), willow, rowan, hazel, aspen, hawthorn, blackthorn and cherry (wild and bird).

Establishment

There are two options for establishing riparian buffers – natural regeneration or new plantings. Natural regeneration is only possible where an appropriate seed source exists. Grazing pressure must be controlled and fencing is not always appropriate, e.g. in areas prone to flooding. If natural regeneration isn't possible, new planting schemes should arrange light-foliaged species in small groups to replicate the vegetation structure of a secondary forest. Minimise machinery use in these riparian areas to reduce damage. Very wet sites should be left unplanted. As with all new plantings, trees may need protection from browsing animals and wildlife, and weed competition controlled, although the use of chemicals in these sensitive areas is not recommended.

Riparian buffers – Management for protection and production

As with windbreaks and hedges, riparian buffers also need some on-going care and management to ensure their effectiveness. The level of management needed will depend on a consideration of site sensitivity, the function of the buffer (e.g. temperature regulation of the water course or reduction of pollution and run-off), its intrinsic value for wildlife and any potential productive function (e.g. for timber or biomass production). Some buffers will benefit from active management including thinning, coppicing or pollarding, whilst others will be more sensitive to the impacts of such management and minimum intervention may be the only option.

Figure 31: **Riparian buffers**

©The Woodland Trust

CASE STUDY: Trees enhance flock health and field drainage

Unable to turn stock into some fields at certain times because of substantial rainfall and lack of shelter, Welsh sheep and beef farmer Jonathan Francis worked with the Woodland Trust to incorporate trees and fencing on his farm to improve shelter, land drainage and grass growing conditions.

In the 2015, Welsh sheep and beef farmer Jonathan Francis planted almost 15,000 trees to help improve the productivity of his 113 ha farm.

Jonathan wanted to address surface water run-off which was affecting the sward, causing waterlogged fields and soil erosion which led to a loss of land alongside water courses. There was also a need for shelter.

Narrow but strategically sited tree belts are very effective at improving field drainage. Research at a group of farms in Pontbren showed that within three years of planting – particularly on a slope – water infiltration rates were improved by 60 times compared to grazed pasture. By increasing soil permeability and water-storing capacity, trees reduce run-off, poaching and consequent damage to the sward. Such improvements also help to reduce flock health issues, such as lameness.

©WTML/Paula Keen

Key facts

- Treed farm boundaries, linear shelterbelts and small clusters of woodland help create sheltered, well-drained fields which provide the best conditions for lambing and good mothering.

- Biosecurity is strengthened, as the potential for disease transmission from neighbouring animals is reduced.

- The risk of neonatal loss of lambs is reduced, and the incidence of mastitis lowered through reduced wind exposure.

- More cost-effective livestock systems, such as outdoor lambing and early turnout, can be practised.

- Reductions in surface run-off and improvements in the land's capability to hold water improve water quality and slow peak flow rates in nearby water courses.

This case study was compiled by the Woodland Trust, for more information see https://www.woodlandtrust.org.uk/publications/2015/06/trees-enhance-flock-health/

Chapter 5
Hedges, windbreaks and riparian buffers

Legal and other considerations

Hedges

A felling licence will be necessary from the Forestry Commission if felling stems that are 15 cm or larger in diameter when measured at breast height (1.3 m from the ground) and if more than 5 m^3 (timber volume) will be felled in any calendar quarter. This reduces to 2 m^3 if any of the wood is to be sold and the licensable diameter reduces to 8 cm or larger if felling single stemmed trees.

If managing by coppicing, find out who owns the hedge before you coppice it, particularly if it is a boundary or roadside hedge. Even if you do own it, you may want to consult your neighbours and inform local residents as coppicing a hedge will have a significant, but temporary, impact on the landscape. Under current (2018) Cross Compliance regulations, hedges and trees can only be trimmed or cut between 1st September and 1st March, although it is possible to carry out hedge and tree coppicing and hedgelaying from 1st March until 30th April. Coppicing in late winter (January/February) allows birds to make good use of the hedgerow berries over the winter.

It is not normally necessary to apply for consent under the Hedgerow Regulations 1997 before coppicing a hedge, provided cut stools are given adequate protection and allowed to regrow. If the intent is not to allow the hedge or any part of the hedge, however small, to regrow then a notice of intent to remove must be submitted to the local planning authority (LPA). You will also need to contact your LPA if any of the trees to be felled or coppiced have a Tree Preservation Order (TPO) or are in a conservation area. Local authorities usually have a map which shows the locations of all TPOs.

Riparian buffers

Before planting riparian buffers, contact your local office of the Environment Agency, Natural Resources Wales or the Scottish Environment Protection Agency to discuss your plans. A flood defence consent may be required for planting close to main rivers (the 'byelaw strip'). Consent may also be needed for any planting within a designated flood storage area. Lead Local Flood Authorities (LLFAs) (or Internal Drainage Boards (IDBs)) are responsible for flood defence consents on ordinary water courses (i.e. river, stream, ditch, drain, cut, dyke, sluice, sewer [other than a public sewer] and passage through which water flows and which does not form part of a main river).

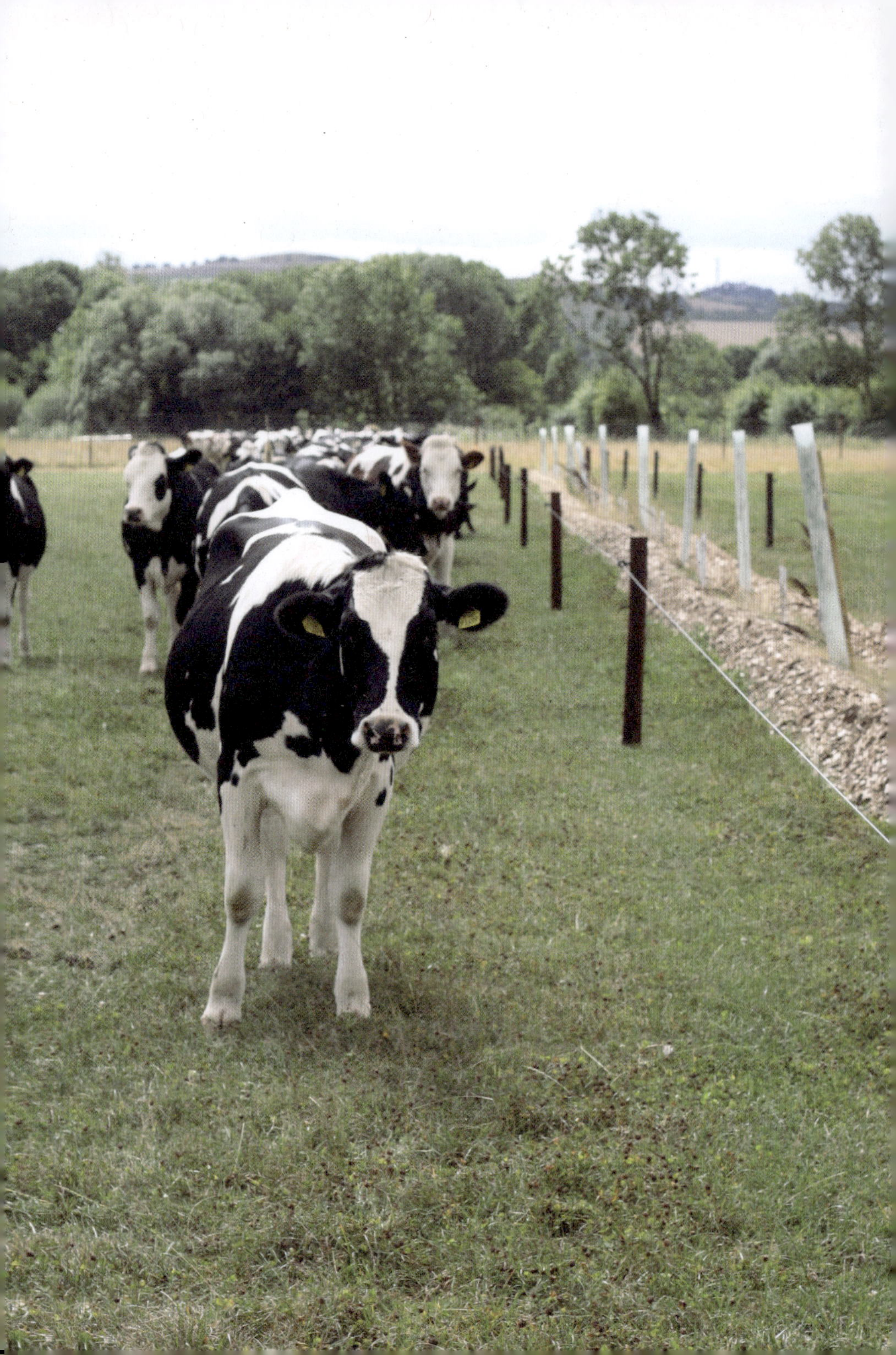

Chapter 6
The economic case for agroforestry

Stephen Briggs and Ian Knight, Abacus Agriculture

Introduction

Research shows that the adoption of agroforestry can increase farm productivity, sustainability, land use efficiency and farm incomes. This chapter covers the economic aspect of agroforestry and the financial principles associated with the economic analysis of four key agroforestry systems: silvoarable, silvohorticulture, silvopasture including lowland and upland silvopasture, with worked examples.

There is a range of potential financial benefits that agroforestry systems can bring, many of which have been listed earlier in the book: reduced stress for plants and animals, increased soil fertility, reduced risk of flooding for instance. The exact figures that would apply to an individual farm or system, however, are almost impossible to predict. Record as much financial detail as you can within the various enterprises to allow you to find out how the trees are affecting your farm as your system develops, as this may allow you to demonstrate a financial gain. As an example your trees may allow earlier and later grazing that won't be attributed to any gross margin calculation on the trees but are positively affecting the bottom line.

This chapter does not claim to give robust financial models for UK agroforestry and should not be used for such. Commercial agroforestry models are still rare in the UK. This chapter aims to give you the general tools and information to allow you to plan an agroforestry system and to assess how it is performing financially. The authors intend to update the handbook as more and better information becomes available.

Agroforestry systems have been shown to be multi-functional, bringing economic, environmental and social benefits to farms. Introducing trees and woody shrubs into an existing crop or livestock system, or livestock into your woodland or forest, typically aims to create financial advantage from additional income and biological interactions.

◀ Dairy stock, electric fencing and woodchip mulch at Eastbrook Farm agroforestry
©Ben Raskin

Chapter 6
The economic case for agroforestry

In the UK, silvoarable systems include the introduction of trees into arable and horticulture systems, where the objectives may include soil protection, diversified value-added income, increased soil organic matter and natural pest and predator populations.

Lowland silvopasture plantings are being trialled to allow cattle to graze tree fodder as self-medication and provide shade and shelter for livestock. In the uplands, high flood risk areas are being planted with agroforestry windbreaks and riparian buffer strips to reduce the impact of heavy rainfall and providing shade and shelter for livestock in extreme weather.

At farm level, the mix of short and long-term multifunctional components including annual crops, livestock and perennial trees can create a management obstacle for farmers considering adopting agroforestry. Often the complexity is a perceived rather than an actual barrier. New agroforestry systems on agricultural land require a longer-term investment to establish viability than short term annual cash cropping.

The economic case for agroforestry can be considered in three main ways:

1. **The value of enhanced ecosystem services from agroforestry systems** – these values may directly enhance the farm enterprise and/or provide wider public benefits. For example, soil improvement and water management might directly benefit the farm enterprise, whilst flood alleviation and biodiversity enhancement are examples of wider public benefits. Others such as carbon sequestration for climate regulation are the subject of current innovation and although clearly an example of an ecosystem service that operates globally as a public good, may offer financial opportunities at a farm enterprise level as well.

2. **The value of enhanced agriculture outputs from agroforestry systems** – these values apply at the farm enterprise level and have been well explored in previous chapters e.g. enhanced grain yield/hectare from silvoarable systems or enhanced meat production from silvopastural systems.

3. **The value that can be derived directly from the tree component of agroforestry systems** – for example the fruit, nuts and berries, timber, woodfuel or direct payment for ecosystem services e.g. carbon finance that are related to the tree component directly

Chapter 6
The economic case for agroforestry

Guide to using the tables in this chapter

Gross margins are widely used across all forms of agriculture. This handbook presents gross margins for agroforestry in a similar way to those commonly used in farm management. However, commonly used gross margins are for a single cropping year, and trees are multi-year and long term. So, for the purposes of providing a snapshot for a farmer to cross compare a single year's cropping, alongside a multi-year tree component, the gross margins in this handbook are for one year of the tree's production cycle, at peak yield. A basic tree/age/yield table is provided underneath each gross margin to show how much to discount yields in relation to the age of the tree.

Implications of agroforestry design on farm economics

Agroforestry systems should be designed to efficiently utilise on-farm resources and provide more output (e.g. fuel, food, carbon capture) than the system consumes as farm inputs (e.g. chemicals, labour, and machinery). Working examples in the UK show that the adoption of agroforestry, when carried out efficiently, with the right design, soils, infrastructure, labour and local markets can thrive with the net benefits outweighing comparable investments in alternative land use.

During the design and planning stage carefully consider the cost implication of tree density, which is determined by the number of trees being planted per unit of land, usually per hectare. The optimum tree density depends on whether the focus is to maximise the original enterprise or to achieve a balanced portfolio of tree, crop and/or livestock outputs. This stage of the planning process is flexible and should be carried out prior to ordering and planting any trees.

Chapter 6
The economic case for agroforestry

Financial evaluation of agroforestry

Most financial analysis is based on a comparison relative to a baseline. In the analyses that we make on the financial aspects of agroforestry, we tend to focus on a marginal cost-benefit analysis. In other words, we look at the things that change. Due to the time periods involved in agroforestry, we need to include labour and machinery costs as these can vary substantially.

Enterprise studies can establish individual gross and net margin contributions to the financial performance of a farm business. To establish the feasibility of a new agroforestry enterprise, assess the possible trade-offs in yield and financial performance of the proposed agroforestry system. Cost accounting for an agroforestry system must consider a range of short- and long-term production cycles for annual and perennial crops, forage, trees and livestock.

The core financial principles are similar to everyday farm budgets and utilise common accounting practices to evaluate the feasibility of agroforestry, these include:

- Whole farm budgeting for profit from agroforestry
- Agroforestry enterprise budgeting
- Gross margin analysis of agroforestry
- Forecasting agroforestry outputs and costs
- Cash-flow forecasting for agroforestry

Farm production from trees differs from annual crop and livestock production cycles as costs (inputs) will potentially be incurred over several years, whilst an agroforestry system establishes and reaches full potential yield and sales (outputs). The period of time required depends on the tree species and end use of the tree product. Apple trees take six years, poplar 15 years, walnut for high-value timber can take up to 60 years to mature.

When undertaking financial forecasting, consider how an annual crop or grazing livestock underneath the agroforestry system may vary over time. For example, year one might be a vegetable cash crop whilst trees are still in their infancy. As the trees mature and need harvesting, an early grass and forage cutting ley might be more suitable to facilitate access in the field for harvest of tree fruit in late summer. Assess the different cropping cycles and management options, to create a real-time picture of your own agroforestry enterprise.

Chapter 6
The economic case for agroforestry

Whole farm budgeting for profit from agroforestry

Use a whole farm budget to show the anticipated financial performance of a farm business over a period of time. In many cases for convenience this is carried out to coincide with the farm's financial year end, traditionally at the end of September (Michaelmas) or the end of March (Lady Day).

The simplest way to create a whole farm budget incorporating agroforestry is to begin with the basics of the existing farming system. How many hectares of crop or head of livestock are to be kept, and how many hectares of agroforestry will be planted? Incorporating how many hectares of agroforestry will be planted versus the change in cropping or stocking area will begin to provide the framework for your whole farm budget incorporating a new agroforestry system.

Using this framework, a detailed picture can be built by estimating the quantities and cost of inputs which each enterprise – crop, livestock and agroforestry – will consume each year. The same estimate is required for the volume of output and prices for each crop, livestock and agroforestry product for the same period, one year. To account for the agroforestry element the annual budget for the existing farm can be duplicated year on year, adjusting the agroforestry element, i.e. tree management, input costs and output sales which are adjusted to show the productivity of the trees as they mature, e.g. fruit trees, year two to four 50% yield, year six to 15 100% yield. Using a spreadsheet can greatly simplify the process.

Adjustments are required to fully reflect the real-time picture of the farm business allowing for opening and closing stocks of crops, livestock and tree produce still on the farm at the beginning of the year and any left on the farm at the end of the financial year. Similarly, any outstanding bills left unpaid (creditors) or sales income not received (debtors) should be included within the whole farm budget.

Adjust the whole farm budget in the first year to deduct any payment or receipt relating to personal, tax or capital. The efficiency of a new agroforestry enterprise should not rely on any personal lifestyle choices nor the settlement of tax obligations, which vary depending on government policy and the skills of your accountant.

Chapter 6
The economic case for agroforestry

The purchase of trees is a capital item lump sum and should be treated the same as the receipt or repayment of loans, the purchase of land, buildings and machinery, and removed from the budget as unrelated to normal trading. Depreciation of wasting assets over the useful life of a machine or implement should be retained in the budget. The resulting whole farm budget (Table 8) shows the successive net profit year on year to gauge the effect of establishing agroforestry within the current farm business scenario.

Figure 32: **Woodchip being processed as additional farm income**

©Ben Raskin

Chapter 6
The economic case for agroforestry

Table 8: Sample trading profit and loss account year ended 31 March 2019–24

Farm income	2019 (£)	2020 (£)	2021 (£)	2022 (£)	2023 (£)	2024 (£)
Sales						
Milk	126,267	138,894	152,783	145,144	137,887	130,992
Calves	10,600	11,660	12,826	12,185	11,575	10,997
Cull cows	9,625	10,588	11,646	11,064	10,511	9,985
Grain	28,431	31,274	34,402	32,681	31,047	29,495
Straw	3,750	4,125	4,538	4,311	4,095	3,890
Agroforestry (apples)						37,240
Basic Payment Scheme	19,800	17,820	16,038	14,434	12,991	11,692
Miscellaneous	2,245	2,470	2,716	2,581	2,452	2,329
Total sales	200,718	216,830	234,949	222,399	210,558	236,620
Less: purchases						
Feed	8,350	9,185	10,104	9,598	9,118	8,662
Livestock	16,986	18,685	20,553	19,525	18,549	17,622
Vet and med	6,317	6,949	7,644	7,261	6,898	6,553
Fertiliser	17,875	19,663	21,629	20,547	19,520	18,544
Seed	5,450	5,995	6,595	6,265	5,952	5,654
Sprays	6,389	7,028	7,731	7,344	6,977	6,628
Agroforestry (running costs)	2,750	3,025	3,328	3,161	3,003	2,853
Regular labour	30,000	33,000	36,300	34,485	32,761	31,123
Machinery (running costs)	22,456	24,702	27,172	25,813	24,523	23,296
Rent and rates	13,750	15,125	16,638	15,806	15,015	14,265
Miscellaneous costs	12,875	14,163	15,579	14,800	14,060	13,357
Bank interest	10,200	11,220	12,342	11,725	11,139	10,582
Machinery depreciation	8,750	9,625	10,588	10,058	9,555	9,077
Total purchases	162,148	178,363	196,199	186,389	177,070	168,216
Net profit	38,570	38,467	38,750	36,010	33,488	68,404

N.B. Agroforestry running costs include pruning, harvesting, understorey weed control

Chapter 6
The economic case for agroforestry

Agroforestry enterprise budgeting

The structure of most agroforestry budgets is to assume values for the amount of product (e.g. tonnes of apple, cubic metres of wood) and values per unit product (e.g. £ per tonnes of apples, £ per cubic metre of wood). Compared to agricultural budgets, obtaining estimates of both the amount and the value of tree products is more difficult. To address this is one of the objectives of producing this handbook.

It is straightforward to work out the *value* of fruit or timber produced but harder to establish the *volume* produced, either in total or more importantly per tree. Rearrange the information from the whole farm budget on an individual basis; list input costs and output sales specific to each part of the farm, so an assessment of profitability can be made for each stand-alone enterprise including a new agroforestry system.

To develop an agroforestry enterprise budget, first create the enterprise budget and then combine these budgets into a cash-flow forecast. Enterprises on the farm are made up from any outputs that generate sales, for instance wheat, milk and fruit or timber from agroforestry. Enterprise budgets allow profitability to be monitored and reported whilst the cash-flow forecast provides a tool to assess the feasibility of an agroforestry enterprise in terms of liquidity (cash) over a period of time, normally one year.

The gross margin system of enterprise budgeting is the easiest way to illustrate the enterprise budget and identifies the individual sales (outputs) and variable costs (inputs); both are easily attributed to an enterprise and vary directly to any minor changes in scale. If cropping is decreased by 5% to account for a 5% increase in agroforestry, then sales and variable costs can be adjusted as a similar proportion for each enterprise.

Step 1: As the production cycle for agroforestry differs and covers a longer period, extra detail is required as to which year sales and variable costs were incurred. As a guide, crop or livestock incomes are listed as sales, as are any output from tree fruit or timber. For example, sales from apple trees will be in year six, walnuts sales may be from year 10 for nuts onward, until the tree is felled for timber 40 years later. Develop a list of all sales under the column of the year the revenue is forecast to be generated.

Step 2: The same list should be developed for variable costs. For an arable enterprise fertiliser, seed and sprays are regarded as variable costs. Typical variable costs for agroforestry would include tree establishment, staking,

Chapter 6
The economic case for agroforestry

guarding, pruning and harvesting. Crucially, add the year in which the variable cost is incurred, for example year one establishment costs, staking guarding, year three pruning, year five fruit spraying, picking. This list should indicate a description of the variable cost, the amount of the cost per tree or per hectare and at which point in time (year) the cost was incurred. To calculate the gross margin, subtract the variable costs (inputs) from the sales (outputs). Remember enterprise gross margins are not an indication of profit.

Figure 33: **Newly planted mixed hedge**

©Jo Smith, Organic Research Centre

Chapter 6
The economic case for agroforestry

Table 9: Sample year 6–15 gross margin from silvoarable agroforestry system

Apple agroforestry

Silvoarable orchard cereal system: 85 trees per hectare with a combine harvester or sprayer boom width of 24–36 metres in between the rows, with tree spacing every 3 m in the row.

Apples – orchard system

Production level	Average per ha	Average per ac	
Yield: tonne/hectare (acre)	1.7	0.7	
	£	£	£/tonne
Output at £900/t	1,530	619	900
Variable Costs £/ha (ac):			
Orchard depreciation	60	24	35
Pruning/clearing	50	20	29
Fertiliser/sprays	81	33	48
Crop sundries	20	8	12
Harvesting	117	47	69
Grading/packing	248	100	146
Storage/bin hire	142	57	83
Packaging	111	45	65
Transport	90	36	53
Commission/levies	115	47	68
Total variable costs	1,034	418	608
Gross margin £/ha (ac)	496	201	292
Silvoarable – top fruit gross margin £/ha (ac)	1,242	503	731

Tree age	Year 1–3	Year 4–5	Year 6–15	Year 16–25
Tree yield	zero	50%	100%	75%

Step 3: When the gross margins for each enterprise are added together to calculate a whole farm gross margin, then fixed costs, labour, machinery and rent can be subtracted to provide a net profit figure for the farm. The final net profit figure should be the same as the previous profit figure defined in the 'Whole farm budgeting' section. The information used in the enterprise budget is identical to the whole farm budget for profit, just the layout has changed.

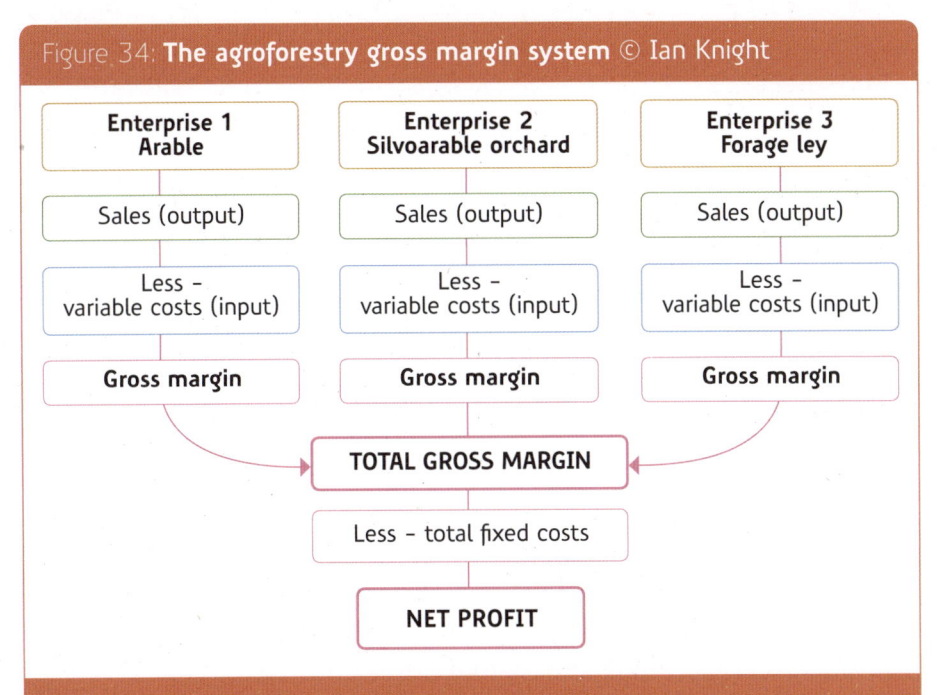

Figure 34: The agroforestry gross margin system © Ian Knight

```
Enterprise 1        Enterprise 2             Enterprise 3
Arable              Silvoarable orchard      Forage ley
   │                     │                        │
Sales (output)      Sales (output)           Sales (output)
   │                     │                        │
Less –              Less –                   Less –
variable costs      variable costs           variable costs
(input)             (input)                  (input)
   │                     │                        │
Gross margin        Gross margin             Gross margin
        ╲                │                   ╱
                 TOTAL GROSS MARGIN
                         │
                 Less – total fixed costs
                         │
                     NET PROFIT
```

This system of enterprise budgeting has the advantage of not allocating fixed costs between individual enterprises. The gross margin system brings simplicity when implementing changes to a farming system such as the establishment of agroforestry. You can evaluate relatively simply the financial effects of planting trees within a field by replacing one gross margin (for instance an annual crop/livestock) with another gross margin for the trees.

Table 10: Sample net revenue increase after replacing 20 ha of cereal crop with 20 ha of agroforestry

		£
Gross margin gain:	20 ha silvoarable agroforestry at £496/ha	9,920
Gross margin lost:	20 ha winter oats at £404/ha	8,080
	Net revenue gain	1,840

Chapter 6
The economic case for agroforestry

For some agroforestry enterprises the gross margin method may not provide sufficient financial analysis; for instance, where pricing for a share or contract farming agreement is concerned. You may need to know exactly how much the agroforestry enterprise will contribute to the farm business and a net margin figure will need identifying. This would involve allocating all costs and overheads that can be easily attributed to individual enterprises, such as a particular labour force or machine, leaving less identifiable general overheads to be allocated to the farm business as a whole. This would provide a net margin profit figure, which would explicitly show whether the share or contract farming agreement would be financially worthwhile.

For small farms without administration staff to manage self-contained enterprises, simply rely on the gross margin cost accounting system, which for most cases remains the most valuable general management accounting procedure, if fixed costs are not ignored.

Agroforestry fixed costs, labour and machinery budgeting

Fixed costs

Most farms already have the equipment, labour and skills for the crop and or livestock component in an agroforestry system, so there are likely to be few changes to fixed costs, labour and machinery for these components. Fixed costs for the tree components are planting costs, pruning and thinning costs, replacing any dead trees (or gapping up) and general management including checking and maintaining tree protection. Fertility or pest and disease management are variable costs. When planning the agroforestry system, the species and spacing should be designed to facilitate easy seasonal management such as pruning and harvesting, to help minimise fixed costs.

Labour

One of the benefits of agroforestry is that often the main labour demand for the tree component is at a time of year when there is less pressure on farm labour. Pruning and tree management is often a late autumn or winter task. Managing tree pests, disease and fertility can typically fit between existing crop and livestock labour requirements. Tree management is slightly less time-critical, often today or tomorrow will suffice. For some farms, agroforestry tree management provides employment activity for staff during quieter seasonal periods.

Machinery

When planning an agroforestry system, ensuring the space between trees (tree groups, alleys, single stands) allows for mechanisation is important.

Aspects to consider are:

- Always work back from the tree and canopy size at or near maturity (trees grow!)
- Consider farm machinery sizes and how multiple machinery widths/passes fit into the spaces between trees e.g. a 24 m working alley will allow the farm to use multiples of 3 m, 4 m, 6 m, 12 m and 24 m width machinery in between the trees.
- Consider 'future-proofing' the design in relation to machinery size expansion e.g. 24 m to 36 m alley widths for increased sprayer capacity or similar.
- In the future, with the potential rise in robotics and small autonomous machines, planning for uniformity of species and spacing between trees may be less of an issue.

Figure 35: **24 m farm sprayer in silvoarable agroforestry in Cambridgeshire**

©Stephen Briggs

Machinery investment

For most farms the machinery investment required for agroforestry is likely to be modest with much of the machinery already present on the farm. Typical investment requirements are given in Table 11.

Table 11: **Sample capital investment costs of tools and machinery**

Item	Typical cost (£)
Chainsaw	250
Hand pruners/saws	40–60
Hand tools (loppers, spades, wire cutters etc.)	20–100
Offset flail mower (understorey management)	5,000
Forklift operator safety cage (for pruning at height)	3,000
Post-hole borer (for planting)	1,700

Gross margin analysis of agroforestry

To develop an agroforestry gross margin:
1. Identify the information required
2. Organise the information for comparative studies.

Estimating the sales (output) of an agroforestry system requires detailed information of the likely value of produce being sold from trees and the financial output from the incorporated farming enterprise in between the trees. Agroforestry combines long-lived perennial trees and bushes with annual crop cycles and so a new approach to gross margin is required to combine the two different time periods. To identify gross profitability of the enterprise as trees mature and start yielding, there will be more than one gross margin for the agroforestry system.

Figure 36: **Wakelyns Agroforestry, Suffolk UK, willow and potatoes**

©Stephen Briggs

Silvoarable gross margins

Below is a sample of physical assumptions, capital investment and gross margins for a top fruit agroforestry system mixed with combinable crop production. Different gross margin analysis is required for different life-cycle periods of the tree component. This approach is required for most agroforestry systems.

Table 12: Sample silvoarable gross margin – wheat and apples

Silvoarable – top fruit

This gross margin sample combines a winter wheat with an apple orchard in year six at 100%. The breakdown of costs are per hectare. In practice this system suits a low-density tree planting of around 85 trees per hectare with a combine harvester or sprayer boom width of 24–36 metres in between the rows, with tree spacing every three m in the row.

Wheat

Feed winter wheat

Production level	Average per ha	Average per ac	
Yield: tonne/hectare (acre)	8.30	3.5	
	£		£/tonne
Output at £150/t	1,245	504	150
Variable costs £/ha (ac):			
Seed	60	24	7
Fertiliser	188	76	23
Sprays	251	102	30
Total variable costs	**499**	**202**	**60**
Gross margin £/ha (ac)	**746**	**302**	**90**

Table continued ▶

Chapter 6
The economic case for agroforestry

Apples – orchard system			
Production level	**Average**		**£/tonne**
	per ha	per ac	
Yield: tonne/hectare (acre)	1.7	1	
	£		**£/tonne**
Output at £900/t	1,530	619	900
Variable Costs £/ha (ac):			
Orchard depreciation	60	24	35
Pruning/clearing	50	20	29
Fertiliser/sprays	81	33	48
Crop sundries	20	8	12
Harvesting	117	47	69
Grading/packing	248	100	146
Storage/bin hire	142	57	83
Packaging	111	45	65
Transport	90	36	53
Commission/levies	115	47	68
Total variable costs	1,034	418	608
Gross margin £/ha (ac)	496	201	292
Silvoarable – top fruit gross margin £/ha (ac)	1,242	503	731

Tree age	Year 1–3	Year 4–5	Year 6–15	Year 16–25
Tree yield	zero	50%	100%	75%

Figure 37: **Silvoarable fruit cereal system in Cambridgeshire**

©Stephen Briggs

Chapter 6
The economic case for agroforestry

Notes and assumptions for Table 12 – silvoarable wheat and apples

- Yields are provided as an average across all winter wheat, all varieties, 1st and subsequent wheats, less 30% for shading trade-off between agroforestry trees and crop.
- Output price as forecast average for 2019. Straw costed as incorporated, returned to the soil as organic matter.
- Wheat seed £380/t C2, sown at 175 kg/ha, costed with single purpose dressing.
- Fertiliser costed as N 22: P 7.8: K 5.6 at £189/t.
- Sprays for wheat costed as annual programme of herbicide (£101), fungicide (£119), insecticide (£8), PGR and other (£24).
- Yield costed from year six–15 at peak output. Yield varies depending on tree age, planting density per hectare, tree spacing in the row, variety, rootstock and growing conditions throughout the season.
- Price costed as an average of all grades sold into the wholesale trade for juicing. Price dependent on variety, grading, end market and buyer.
- Variable costs taken as average between low production level dessert and culinary apples planted at a tree density of 85 trees per ha.
- Orchard depreciation covers capital cost of apple orchard establishment including land preparation, trees, stakes, ties and planting.
- Fertilisers, sprays, predators – a relatively small proportion of gross margin to cover the cost of crop nutrition, minor fungicide applications and predator control.
- Crop sundries includes tree ties, rabbit guards, replacement stakes, tree replacement, bee hire, picking hods, bin depreciation.
- Harvested on an average labour cost £69/t which includes management oversight, National Insurance and allowance of staff holidays. Likely to vary significantly in practice and dependent on variety, yield, fruit size and quality.
- Apples grading and packing labour costed as £146/t, in practice this varies considerably, dependent on working conditions, packing line equipment, staff and crop quality.
- Storage bin hire assumed at £25/300kg bin.
- Packaging costed as an average of between £55 and £167/t. In practice this cost will vary depending on equipment available at the farm or whether using specialist hire. The pack size also determines the packaging price range.
- Transport is costed as travel between farm and pack house and on to the final customer.
- Commission/levies covers marketer and retailer commission as well as AHDB English Apples and Pears Levy payment.

Silvohorticulture gross margins

Below is a sample of physical assumptions, capital investment and gross margins for a short rotation coppice agroforestry system mixed with brassica crop production. You would of course not grow brassicas in the same field in successive years, so in practice you would have a range of different crop gross margins that would sit alongside the tree one. Some horticultural crops would suit the first year after planting while others will cope with greater shade as the trees develop.

Table 13: Sample silvohorticulture gross margin – calabrese and sweet chestnut

Silvohorticulture – sweet chestnut

This sample of a silvohorticultural gross margin outlines a field-scale vegetable crop of broccoli being grown in 24 m alleys within a sweet chestnut plantation, with a tree density of 90 trees per hectare.

Calabrese (broccoli)
One-year field scale vegetable crop

Production level	Average per ha	Average per ac	
Yield: tonne/hectare (acre)	10	4	
	£	£	£/tonne
Output at £422/t	4,220	1,708	422
Variable costs £/ha (ac):			
Seed	907	367	91
Fertiliser	400	162	40
Sprays	265	107	27
Casual labour	1,605	650	161
Packaging/consumables	269	109	27
Levy	20	8	2
Total variable costs	3,466	1,403	348
Gross margin £/ha (ac)	774	313	77

Chapter 6
The economic case for agroforestry

Sweet chestnut agroforestry
Sweet chestnut – orchard system

Production level	Average		
	per ha	per ac	
Yield: tonne/hectare (acre)	0.45	0.18	
	£	£	£/tonne
Output at £2,650/t	1,193	483	2,650
Variable Costs £/ha (ac):			
Orchard depreciation	108	44	240
Pruning/clearing	44	18	98
Fertiliser/sprays	72	29	160
Crop sundries	8	3	18
Harvesting	87	35	193
Grading/packing	108	44	240
Packaging	72	29	160
Transport	81	33	180
Commission/levies	19	8	42
Total variable costs	599	242	1,331
Gross margin £/ha (ac)	594	240	1,319
Silvohorticulture – sweet chestnut gross margin £/ha (ac)	1,368	554	3,039

Tree age	Year 1–5	Year 6–13	Year 14 onwards
Tree yield	zero	50%	100%

Notes and assumptions for Table 13 – silvohorticulture calabrese and sweet chestnut

- High-quality land/soils are required to grow field-scale vegetables. The overheads of production are significant, and a market is required in advance of cultivation. The unit of sale can vary depending on the arrangement between the retailer and grower. This is a general output example for wholesale calabrese.
- Working capital levels identified in variable costs are high relative to other cropping systems and there are significant commercial risks associated with crop quality and demands of the market.
- A range of inputs may be required such as mineral fertilisers and

Chapter 6
The economic case for agroforestry

- lime. Some high organic-matters soils may require manganese supplement.
- We have assumed the production system is organic.
- Harvesting is carried out by hand on picking rigs for pre-pack markets.
- Grading cost includes grading for wholesale/pre-packed packaging at pack house.
- Yield example costed as year 15 at peak yield of 25kg/tree less 15% for grading waste losses. Yield varies significantly with tree density, age and variety, this example is costed as 25kg/tree x 18 trees/ha.
- Output highly dependent on crop quality and grading. Example pricing taken as a graded average between Grade 1 £4,200 and Grade 2 £1,100.
- Orchard depreciation covers capital cost of sweet chestnut orchard establishment including land preparation, trees, stakes, ties and planting.
- Pruning and clearing includes tree shaping, ground clearance of brush and mowing 2.92 ha at £12/ha three times a year
- Fertiliser and sprays include foliar feeds and organic fertilisers, pelleted chicken manure, mineral fertilisers.
- Crop sundries include tree ties, rabbit guards, replacement stakes, tree replacement, bee hire, picking equipment, bin depreciation.
- Transport is costed as travel between farm and pack house and on to the final customer.
- Commission/levies covers marketer and retailer commission.

Figure 38: **Brassicas and apple trees at Duchy Home Farm**

©Ben Raskin

Chapter 6
The economic case for agroforestry

Lowland silvopasture gross margins

Below is a sample of physical assumptions, capital investment and gross margins for a lowland silvopasture agroforestry system.

Table 14: Sample gross margin lowland silvopasture – walnut and clover ley silage

Silvopasture – walnuts

This gross margin example outlines a five-year cutting ley with walnut trees planted at 28 m between the rows and 3 m tree spacings within the row, which equates to 27 trees per hectare.

Typically, this field would be an arable field or temporary grass field on a livestock farm which forms part of a wider rotation, with the five-year cutting ley providing a cash crop of big bale silage for selling off farm. This silvopasture system can be described as a walnut orchard. This system could use early and late-pollinating varieties such as Broadview, Buccaneer, Frankette, Rita, Northdown Clawnut.

Clover ley – silage

Five-year grass and white clover cutting ley

Production level	Average		
	per ha	per ac	
Yield: tonne/hectare (acre)	28	11	
	£		£/tonne
Output at £50/t at 30% dry matter big bales	1,400	567	50
Variable Costs £/ha (ac):			
Establishment and seed	198	80	7
Fertiliser	41	17	1
Sprays	19	8	0.7
Silage conservation costs £/ha:			
Silage additive and wrap	73	30	3
Contractor charges	261	106	9
Net variable costs five-year white clover ley	91	37	3
Total variable costs	**683**	**276**	**24**
Gross margin £/ha (ac)	**717**	**290**	**26**

Table continued ▶

Chapter 6 — The economic case for agroforestry

Walnut agroforestry
Walnut – orchard system

Production level	Average per ha	Average per ac	
Yield: tonne/hectare (acre)	0.39	0.16	
	£		£/tonne
Output at £3,100/t	1,209	489	3,100
Variable costs £/ha (ac):			
Orchard depreciation	162	66	415
Pruning/clearing	66	27	169
Fertiliser/sprays	108	44	277
Crop sundries	11	4	28
Harvesting	130	53	333
Grading/packing	162	66	415
Packaging	108	44	277
Transport	81	33	1
Commission/levies	19	8	49
Total variable costs	847	343	2,172
Gross margin £/ha (ac)	362	146	928
Silvopasture – walnut gross margin £/ha (ac)	1,079	437	2,767

Tree age	Year 1–5	Year 6–13	Year 14 onwards
Tree yield	zero	50%	100%

Notes and assumptions for Table 14 – lowland silvopasture – walnut and clover ley silage

- Silage yield costed at 14t/ha fresh weight, taken over two cuts at 30% DM, possibly grazing aftermath, herbage yields dependent on site class and rainfall.
- Silage output costed as £20/bale with each bale weighing approx. 400kg.
- Establishment costs vary depending on method and equipment/contractors' costs, Seed prices based on typical merchant mixtures.
- Fertilisers are applied rotationally based on soil analysis.
- Sprays are often unnecessary except in conventional systems with heavy broadleaved weed population early in the season.
- Silage additive and wrap is costed as an average for big-bale wrapping and additive.

Chapter 6
The economic case for agroforestry

- Contractor charges cover two cuts of big bale silage at 14t/ha fresh, carted and stacked in the yard.
- Net variable costs cover for five-year cutting ley.
- Walnut yield example costed as year 15 at peak yield of 20kg/tree less 15% for grading waste losses. Yield varies significantly with tree density, age and variety, this example is costed as 5kg/tree x 27 trees/ha.
- Walnut output highly dependent on crop quality and grading. Example pricing taken as a graded average between Grade 1 £5,000 and Grade 2 £1,300.
- Orchard depreciation covers capital cost of walnut orchard establishment including land preparation, trees, stakes, ties and planting.
- Pruning and clearing includes tree shaping, ground clearance of brush and mowing 2.92 ha at £12/ha three times a year.
- Fertiliser and sprays for walnuts include foliar feeds and organic fertilisers, pelleted chicken manure, mineral fertilisers.
- Crop sundries includes tree ties, rabbit guards, replacement stakes, tree replacement, bee hire, picking hods, bin depreciation.
- Transport is costed as travel between farm and pack house and on to the final customer.
- Commission/levies covers marketer and retailer commission.

Figure 39: **Upland silvopasture system in Scotland**

©Stephen Briggs

Forecasting agroforestry outputs and costs

Budgeting is only of any value if the information recorded is accurate. Collecting quality information to estimate tree produce quantities and prices for fruit or timber will provide the foundation to forecast sales output and variable cost input values.

Commonly used information relating to livestock and arable costs and prices are regularly taken from the John Nix Farm Management Pocketbook[1] and the Organic Farm Management Handbook[2] which both provide an up-to-date-accurate resource for building whole farm and enterprise budgets. Budgeting for agroforestry, however, requires additional market research using your own experience of how trees grow on your farm and the knowledge of industry professionals from farming and forestry sectors. Tree nurseries and tree suppliers are another useful resource to build tree volume data, and fresh produce and timber merchants are often pleased to help a new agroforestry enterprise.

Impact on outputs from tree and crop competition

Combining productive agroforestry components, trees and crops or trees and pasture results in challenges for budgeting and creating gross margins, especially in accommodating competition between the crop and tree components, and dealing with annual and perennial growth cycles that change in size, productivity and competition over time.

In newly planted agroforestry systems, when trees are small, there is little competition to the grass or crop component between and around the trees. There is therefore no need initially to discount or adjust crop or grass yield per unit area. As the trees grow there will be more competition for water, nutrient use and shading, depending on how the tree canopy is managed. Older, larger trees can have a negative impact on crops and grass productivity near the tree base, under the canopy or shaded area. This negative impact on crop yield reduces with the distance from the tree. Managing the tree crown by regular pruning can minimise the competition to the understorey crop and facilitates access for machinery.

This variable competition and associated 'crop loss' should be balanced by the benefits that trees will bring to your system. Initially these benefits will be small but as the trees grow the benefits will increase, inversely proportional to the competition. Figure 40 (Graves et al. 2007)[4] demonstrates this, but shows that the overall yield of the two elements combined is almost always more than either of them grown as a monoculture.

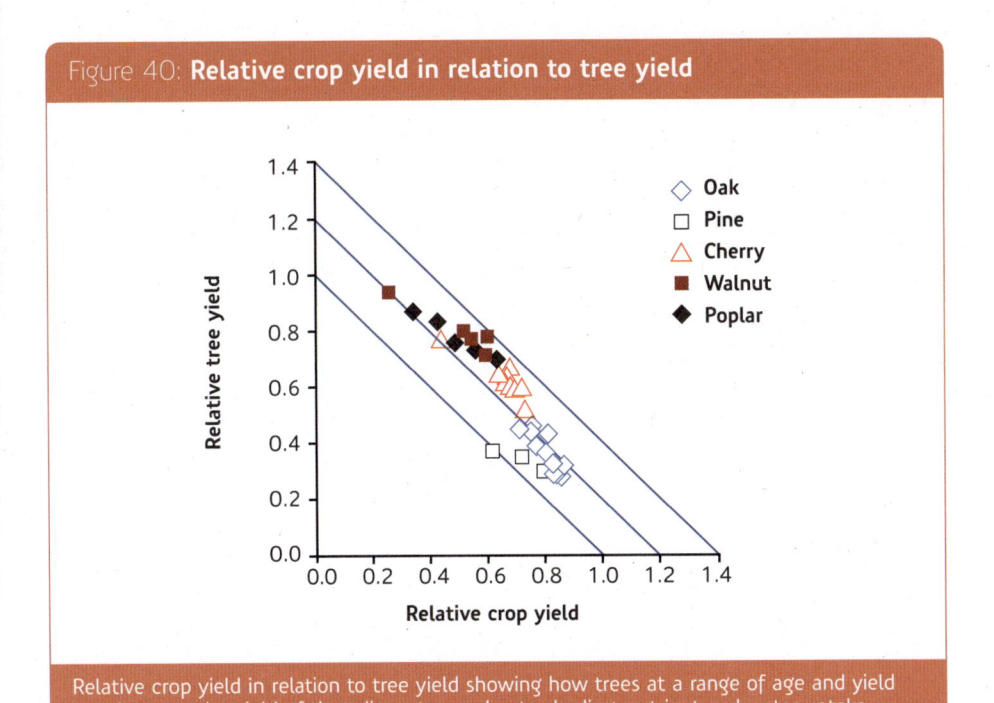

Figure 40: Relative crop yield in relation to tree yield

Relative crop yield in relation to tree yield showing how trees at a range of age and yield can decrease the yield of the adjacent crop due to shading, nutrient and water uptake

Copyright: Graves, A.R., Burgess, P.J., Palma, J.H.N., Herzog, F., Moreno, G., Bertomeu, M., Dupraz, C., Liagre, F., Keesman, K., van der Werf, W. Koeffeman de Nooy, A. & van den Briel, J.P. (2007). Development and application of bio-economic modelling to compare silvoarable, arable and forestry systems in three European countries. Ecological Engineering 29: 434-449

Chapter 6
The economic case for agroforestry

Comparing how trees grow in woodland and agroforestry systems

The challenge for budgeting and creating gross margins on an annual basis is that trees grow differently to crops and growth potential also differs in an agroforestry system versus a forestry plantation. Work at the Agri Food and Biosciences Institute (AFBI) in Northern Ireland compared tree heights and trunk diameters of different trees grown as woodland and as agroforestry at their Loughgall site (McAdam, J. et al).

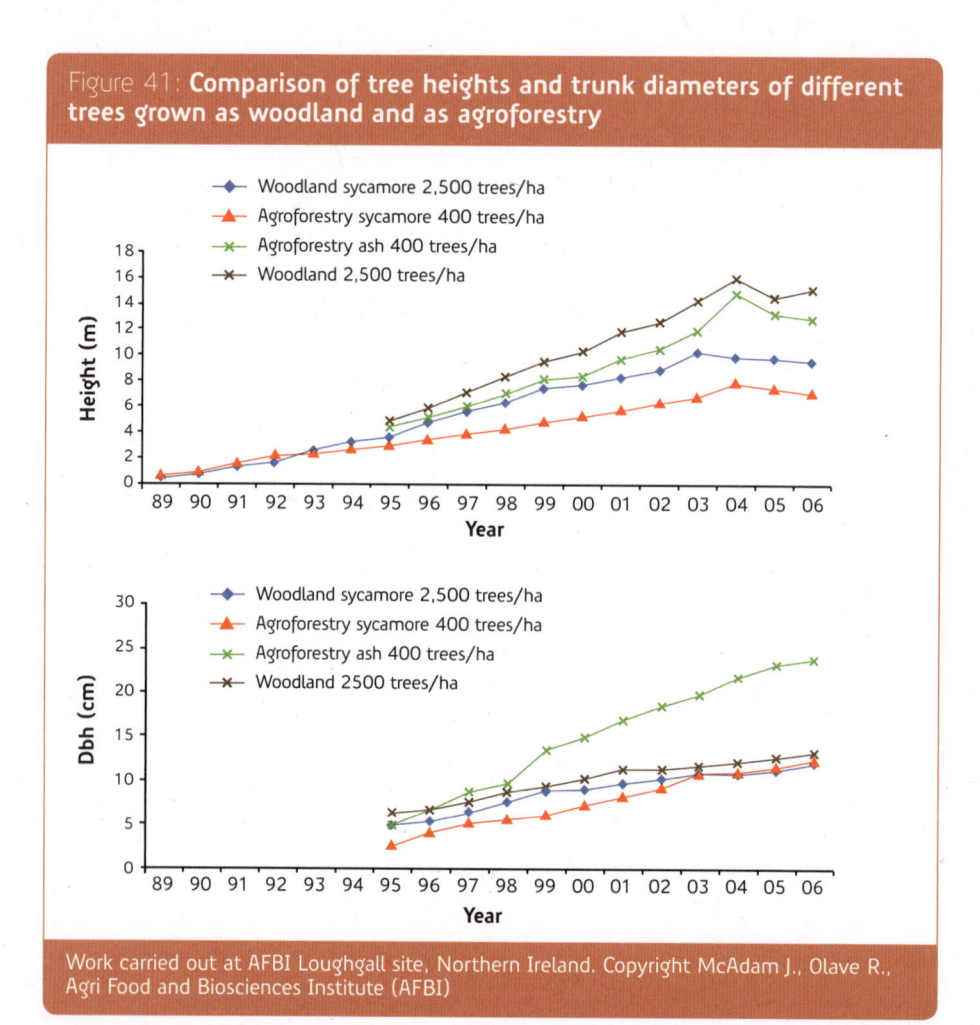

Figure 41: **Comparison of tree heights and trunk diameters of different trees grown as woodland and as agroforestry**

Work carried out at AFBI Loughgall site, Northern Ireland. Copyright McAdam J., Olave R., Agri Food and Biosciences Institute (AFBI)

Research at Loughgall found that both ash and sycamore trees grow taller in woodland settings compared to agroforestry over the same time frame. In woodland they compete with each other for light, which forces height growth. Conversely, ash and sycamore trees grown as agroforestry tend to grow larger-diameter trunks, especially so for ash species, over the same time period.

With less intra-competition between trees afforded by the agroforestry system, trees tend to grow more quickly but not as tall, with a larger-diameter trunk girth. These productivity characteristics need to be considered when creating or modifying gross margins.

Comparative productivity in upland systems

Agroforestry can be used in the upland environment and provides many benefits in terms of shelter for livestock, soil and water management, and landscape value. Research on upland silvopastoral agroforestry systems undertaken at the Macaulay Land Use Research Institute[3] has shown that:

- Trees can be planted at wide spacing within upland sheep-grazed pastures with no reduction in agricultural productivity for up to 10 years with fast-growing tree species (for example, hybrid larch, Figure 42) and for at least 12 years with slower-growing tree species (for example sycamore and Scots pine).

- Pruning of trees reduces shading of the pastures and can result in high levels of agricultural productivity even after 10 or 12 years (Figure 42). Pruning will also increase the quality of the timber grown.

Figure 42 shows that for the first seven years of an agroforestry system, as the trees grow taller, annual agricultural production was unchanged. Productivity increased to year nine over two dry summers. Year 10 was a wet summer with no advantage to agroforestry. In year 11 the tree canopies were big enough to shade the pasture significantly. After that, pruning of the trees held agricultural production at 90% of conventional pasture output.

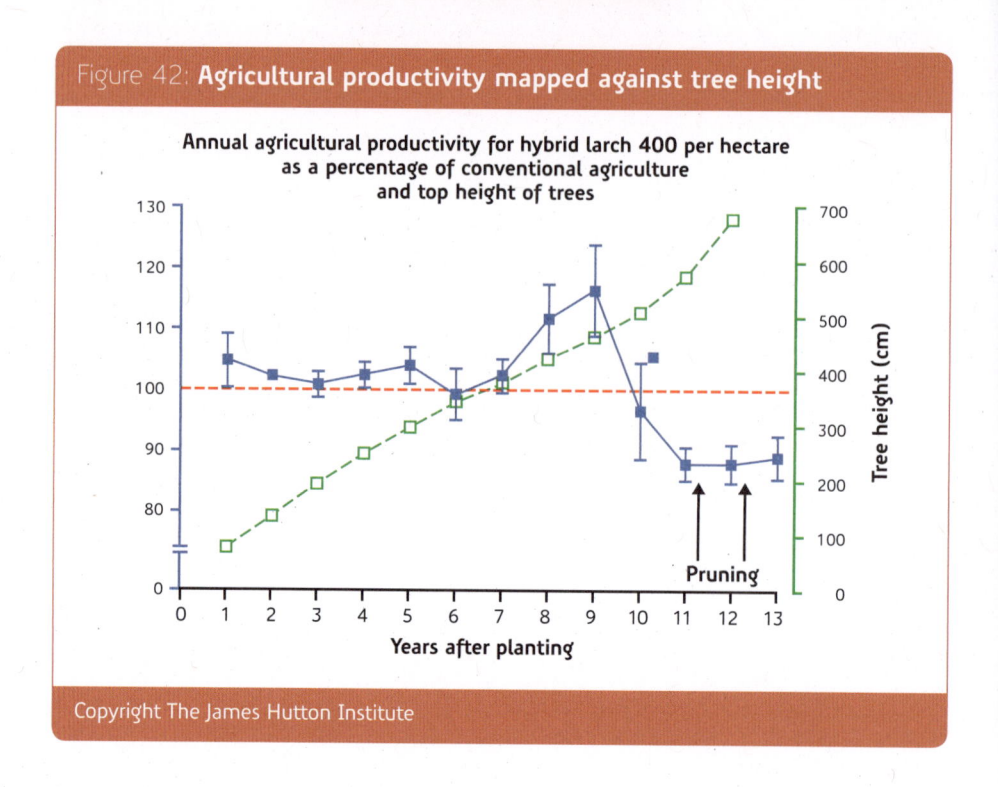

Figure 42: Agricultural productivity mapped against tree height

Annual agricultural productivity for hybrid larch 400 per hectare as a percentage of conventional agriculture and top height of trees

Copyright The James Hutton Institute

Productive potential of agroforestry – Land Equivalent Ratio

The easiest way to explain the productive potential of agroforestry is by using a Land Equivalent Ratio (LER) calculation. As described on page 25, LER is a way of comparing the combine yield potential of mixed crops compared to monocultures of each of them. A pan-European project SAFE[5] looked at tree and crop yields for 42 tree-crop combinations and modelled LERs of at worst 1.0 (the same as monoculture) and at best 1.4 (40% more productive). Most LERs reported were in the range 1.2–1.3, i.e. agroforestry being 20–30% more productive than monoculture farming systems.

Examples of Land Equivalent Ratio from different agroforestry systems

Apple and wheat

Table 15 shows an example of a silvoarable agroforestry system combining cereal production (wheat) with fruit trees (apple) to evaluate productive potential and estimate an LER. This assumes the fruit trees use 8%, and the cereal component 92% of the land area and that the relatively small fruit trees do not negatively impact on cereal productivity, with only minimal shading, water and nutrient competition.

Table 15: Year three sample LER spreadsheet calculation – apple/wheat

Apple / wheat agroforestry	Year three				
	Land area %	Yield ha/yr	Value £/t	Component output £/ha/yr	Total output £/ha/yr
Monoculture					
Apple orchard @ 1,000 trees/ha	100	10.4 t	650	6,760	**6,760**
Wheat	100	10 t	150	1,500	**1,500**
Agroforestry					
Apple @ 90 trees/ha	8	1.1 t	650	715	715
Wheat	92	9.5 t	150	1311	1,311
					2,026
LER = 1.06	1.1	Tree agroforestry yield / Tree monoculture yield	+	Crop or livestock agroforestry yield / Crop or livestock monoculture yield	9.5
	10.4				10

In years nought to three, zero productivity from the fruit trees is assumed and there is a negative impact on the output and enterprise budget. From year three a modest fruit yield is projected with a total economic output slightly greater than a wheat monoculture but less than a full orchard. This creates a Land Equivalent Ratio (LER) of 1.06.

From year eight when the fruit is in full production a total economic output is far greater than a wheat monoculture and is projected only slightly lower than a full orchard. This creates a Land Equivalent Ratio (LER) of 1.4.

Table 16: **Year eight sample LER spreadsheet calculation – apple/wheat**

Apple / wheat agroforestry	Year eight				
	Land area %	Yield ha/yr	Value £/t	Component output £/ha/yr	Total output £/ha/yr
Monoculture					
Apple orchard @ 1,000 trees/ha	100	10.4 t	650	6,760	**6,760**
Wheat	100	10 t	150	1,500	**1,500**
Agroforestry					
Apple @ 90 trees/ha	8	6.00 t	650	3,900	3,900
Wheat	92	9.0 t	150	1,248.9	1,248.9
		1.0t/ha reduced from shading			
					5,149
LER = 1.4	6	$\frac{\text{Tree agroforestry yield}}{\text{Tree monoculture yield}} + \frac{\text{Crop or livestock agroforestry yield}}{\text{Crop or livestock monoculture yield}}$			9
	10.4				10

When tree components are used in agroforestry that compete more with the adjacent crops of cereals, vegetables and grass, the productivity of the adjacent components should be discounted accordingly.

Cereal and short rotation coppice

Table 17 is an example of a silvoarable agroforestry system combining cereal production (wheat) with short rotation coppice (SRC) to evaluate productive potential and estimate an LER. This assumes that the SRC willow uses 20% of the land area and the wheat 80% of the land area and that the SRC willow impacts on c.50% of the wheat productivity through shading, water and nutrient competition.

Chapter 6
The economic case for agroforestry

Table 17: Sample LER spreadsheet calculation – short rotation coppice, willow/wheat

Short rotation coppice (SRC) willow / wheat agroforestry

	Land area %	Yield ha/yr	Value £/t	Component output £/ha/yr	Total output £/ha/yr
Monoculture					
SRT plantation willow	100	8.33 odt	60	499.8	**499.8**
Organic wheat	100	10 t	150	1,500	**1,500**
Agroforestry					
Willow	20	3.35 odt	60	201	201
Wheat 100%	67	9 t	150	1,350	
Shaded wheat 50%	13	0.9 t	150	135	
		9.9 t	150		1,485
					1,686
LER = 1.39		3.35	Tree agroforestry yield / Tree monoculture yield + Crop or livestock agroforestry yield / Crop or livestock monoculture yield		9.9
		8.33			10

The productive and financial potential of a willow short rotation coppice and wheat agroforestry system is greater than that of a monoculture system.

Cash-flow forecasting for agroforestry

The objective of an agroforestry cash-flow budget is to indicate the flow of cash in and out of the farm business over an extended period – between planting trees whilst establishing the enterprise, to harvesting the first productive crop from mature trees.

Conventional cash-flow budgets typically cover a period of 12 months for a standard financial year. Agroforestry requires the incorporation of a longer planning period effectively linking together a series of annual gross margins for understorey crop/livestock enterprises with a series of tree-based gross margins for specific time periods. This allows the cash-flow budget to show the flow of funds, allowing for additional income and expenditure associated with trees.

Chapter 6
The economic case for agroforestry

Subtracting the total cash payments from the total cash receipts for each agroforestry system over a 12-month period provides an indication of the net cash flow for each year. When the opening bank balance from the beginning of the year and the net cash flow are added together a closing bank balance for the year end is shown by the cash-flow budget. This should be carried out for every year in the run-up to maturity of the agroforestry system, when sales from tree products have reached the first stage of full productivity. The agroforestry cash flow will then show the effect that the longer-term investment in trees has on the farm business cash flow, illustrating a positive or negative closing bank-balance figure.

For each budget year in the run-up to the forecast productive year (e.g. year five for apples, year 12 for poplar, year 50 for walnut) make a list of all cash receipts and cash payments likely to affect the farm business bank balance in each financial year.

The profit budget, especially if this is in gross margin form, provides a good basis on which to compile a list of cash receipts and cash payments. It is crucial to deal only with cash items. Personal payments or receipts of tax or capital should be accounted for within the cash-flow budget. Debtors, creditors, stock valuations and depreciation should not be included in the cash-flow budget.

As the process of cash-flow budgeting for agroforestry is carried out over a longer period, a farm business should carry on with the conventional principle of annual cash-flow budgeting but have the longer cash-flow budget of, for example, five, 10 or 50 years, as a separate appendix to the main farm cash flow for annual cropping and livestock. The five-year average for the main farm cash flow end-of-year results should feed into the annual agroforestry cash-flow.

After the detailed list of cash receipts and cash payments has been completed the pattern of timing by which these cash amounts flow in and out of the farm business bank balance can be estimated by examining each payment or receipt in turn. Decide when and how each receipt or payment will be split between one year to the next. This will identify the annual cash flow relevant from the main farm cash flow, giving an opportunity to consolidate these into the opening and closing balances for the annual agroforestry cash-flow budget.

These estimates will need to consider the physical cropping or livestock production cycle on the farm, such as cereal harvest/sales or calving date and pattern of milk receipts. Then any opening or closing payments can be rolled over into the current or following year. It is standard practice to give a four-week credit allowance on sales, and cash receipts should be expected to be paid a month after invoicing. This is an important factor to consider allowing payments at the year start and end to be included in the correct year of the agroforestry annual cash-flow budget. Payments from opening debtors should be consolidated into the final annual figure.

This procedure, of making the credit allowance, should be repeated for each year of the agroforestry planning period. There may also be cases, for bigger purchases such as fertilisers or machinery where a lengthy credit allowance of three months is applicable. These allowances are important in creating a realistic picture of annual cash flows through the farm business.

The final agroforestry cash-flow budget can now be finalised by entering the cash receipts and payments into the main body of the worksheet table according to the decided timings, and calculating the annual opening and closing bank balances. Start with the first-year column, total the annual cash receipts then subtract the annual cash payments made for each individual enterprise, e.g. winter wheat, grass ley, trees as agroforestry, to give the year's net cash flow. This figure is added to the opening bank balance with the resulting amount illustrating the closing bank balance for the year – a positive figure illustrates the bank has money in a credit balance, a negative figure illustrates an overdrawn balance.

Setting up any new enterprise can create an allocation of cash funds into fixed or long-term wasting assets – the risks associated with borrowing should be qualified as to how long any overdraft (if any) requires maintenance before a new agroforestry enterprise is embarked on.

The first year's closing bank balance becomes the second year's opening bank balance and the above process is repeated for each of the remaining years of the agroforestry cash-flow budget-planning period. If an overdraft position is reflected as a negative closing bank balance in any one year, this can create a cash-flow budget accounting problem as interest is charged monthly. It would be better to estimate any overdraft interest payable within the main annual farm budgets and transfer a consolidated overdraft interest figure into the agroforestry cash-flow budget – the result can be entered into the cash-flow worksheet table as overdraft interest under the payments heading.

Your cash-flow chart should illustrate the changes in the annual closing bank balance taken from the agroforestry cash-flow budget, providing a useful tool to assess the impact on cash funds of establishing an agroforestry enterprise on the farm.

Agroforestry cash-flow budget assessment

Maintaining a positive cash position within a business is key to facilitate trading. Cash-flow budget assessment and the implications agroforestry has on the existing farm business are critical to the success of agroforestry over the long term. The objective of the agroforestry cash-flow budget is to illustrate the impact that the long-term costs associated with planting and maintaining trees will have on the farm business. The main test of this objective is the maximum tolerable overdraft, unless the farm has sufficient cash (liquidity) to service the new agroforestry enterprise without the bank balance entering a negative (borrowing requirement) position. The next test is to assess if the cash-flow budget shows a consistent deficit of cash at bank as a negative balance. An important question to raise at the point of time where the bank balance shows a negative balance is whether this negative balance exceeds the agreed terms of any borrowing provided. The next assessment should be to identify any significant variations in bank balances from one year to the next. What was the key item responsible for this? Was it tree-related or a fluctuation in commodity price on the farm? What strategies are required to manage significant variations in cost?

These checks may show problems, such as a lack of affordability, or negative impacts on the cash trading balance. There are strategies to assess the impact of change on the farm business and what needs to happen before the adoption of a new agroforestry enterprise takes place. There are methods of increasing cash flow through the farm business whilst the agroforestry enterprise matures and becomes productive. This may mean reducing capital expenditure on trees by adjusting the design or density, or delaying additional capital expenditure on machinery purchases. There may be scope for selling redundant machines after a change in farming practice. Private drawings can be postponed, or general overheads could be reduced. A closer analysis of gross margins could flag an opportunity to increase output or reduce variable costs, although this should have been appraised and actioned as part of the profit-budget review process.

Land tenure

Agroforestry is an easier decision where the land is owned, and long-term management and land-use decisions can be made. Where land is rented, this adds a layer of complexity to the development of an agroforestry system. For longer-term tenancies of 20+ years, agroforestry's many tree components can reach a harvestable size and quality and, as such, can be treated like any other farm enterprise. For shorter-term tenancies of three, five or even 10 years, the development of agroforestry can at first sight seem impractical. However, through discussion with landowners there are a number of potential opportunities:

- On rented land, use shorter life-span rotation trees and shrubs, or species that provide shorter-term products, i.e. fruit, berries, nuts, resins.

- Capital expenditure on trees funded by the landlord who owns the tree as a capital asset (as with a building). The tenant farmer manages the trees and receives a proportion of the income from any timber or tree produce in return for management (the landlord also receives a proportion of the income) during the period of their tenure.

- Develop a joint venture between landlord and multiple tenants, whereby one tenant manages the land between the trees with crops/animals and (an)other tenant(s) manage(s) the trees and their harvest. This allows different persons with different skill-sets to manage components that play to their strengths.

To encourage tree planting on tenanted farms in the future requires creative, joined-up thinking by both parties, the landlord and the tenant. Modern farm tenancies and share farming agreements that allow extended tenancy terms, as well as shared cost/income partnerships can be drafted in order to develop tenancy models to suit any individual circumstance. There are good examples of this: such as at Whitehall Farm[6], where Farm Business Tenancies were renegotiated to allow trees to established and yield sufficient for the tenant farmer to show a capital return on the investment in agroforestry; or at The Dartington Trust[7], where multiple enterprises and ownership of tree enterprises within a field has encouraged two or three small-scale producers to produce a range of tree and forage crops together in one large field parcel.

Chapter 6
The economic case for agroforestry

Government support for agroforestry

Whilst most agricultural policy and rules of governance in farming have been agreed at EU level, competency in forestry and woodland has remained a member state issue. In some countries, this has led to a separation of land uses and barriers to practices such as agroforestry that cross the divide, causing confusion, contradictions and complications on the ground.
The position and definition of agroforestry is currently unclear within UK governments, with divergent approaches to integrating trees and farming across the Devolved Administrations. In Scotland, for example, the minimum tree density under the current agroforestry scheme is 200 trees/ha, meaning that farmers in receipt of agroforestry payments are excluded from the often more lucrative Common Agricultural Policy (CAP) Basic Payment Scheme (BPS).

Sitting in this policy and delivery void between forestry, environmental stewardship and agriculture, agroforestry has struggled to thrive in the UK. This uncertainty has passed through the entire sector, confusing landlords, agents and advisors, foresters and farmers alike. Progress has been made recently (2018) in England, where BPS was reformed to allow trees on farms so long as agricultural activity can continue, but such a technical change does little to inform and engage the farming community in the possibility of agroforestry.

Figure 43: **Mixed trees and sheep at the Allerton Project in Leicestershire**

This trial set up by the Game & Wildlife Conservation Trust and the Woodland Trust is testing whether agricultural activity can continue within a range of tree planting densities[8]

©Jo Smith, Organic Research Centre

Chapter 6
The economic case for agroforestry

Agroforestry for carbon capture

There is significant potential for using agroforestry systems to build carbon in woody components and sequester carbon in the soil. Research has demonstrated carbon sequestration of between 1.0 to 4.0 tonnes of carbon per ha per year from agroforestry tree densities of between 50–100 trees per ha. Faster-growing trees with higher density sequester more than slower-growing less densely populated systems as shown in Figure 44.

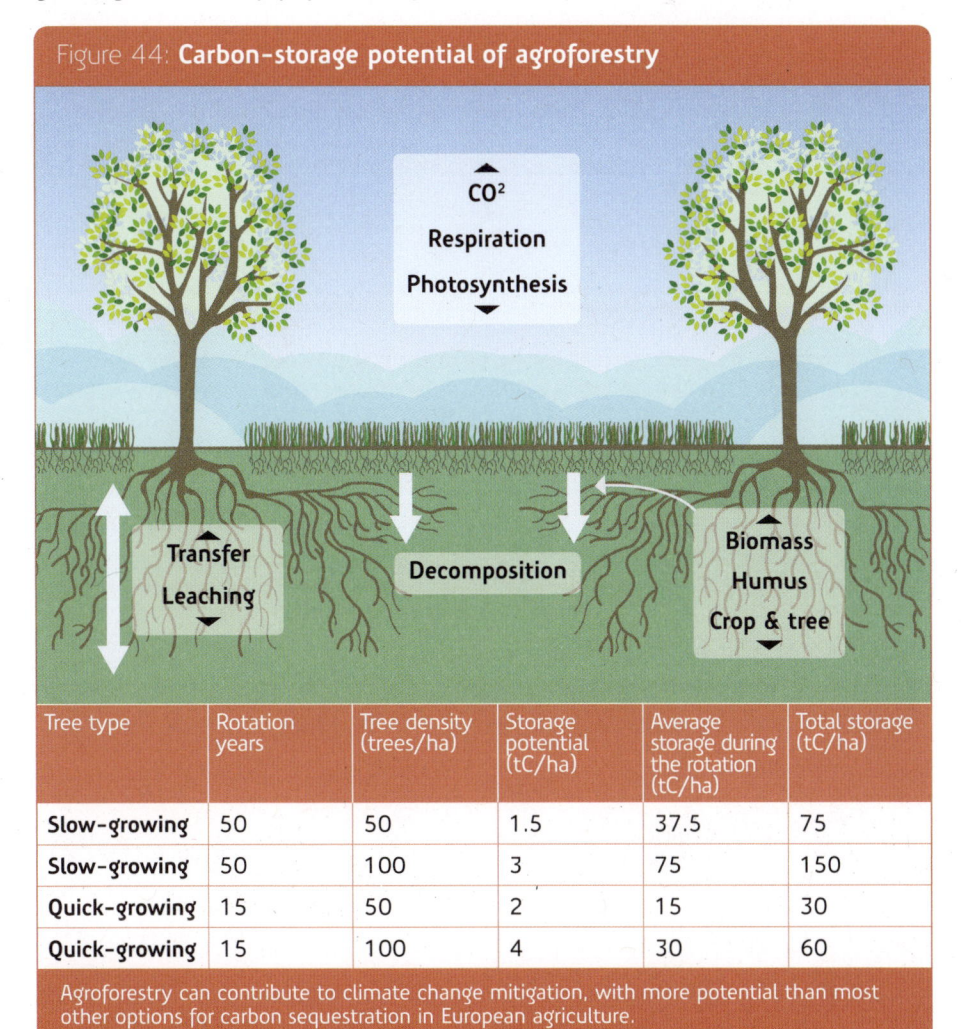

Figure 44: **Carbon-storage potential of agroforestry**

Tree type	Rotation years	Tree density (trees/ha)	Storage potential (tC/ha)	Average storage during the rotation (tC/ha)	Total storage (tC/ha)
Slow-growing	50	50	1.5	37.5	75
Slow-growing	50	100	3	75	150
Quick-growing	15	50	2	15	30
Quick-growing	15	100	4	30	60

Agroforestry can contribute to climate change mitigation, with more potential than most other options for carbon sequestration in European agriculture.
Copyright Dupraz, C. INRA/EURAF

Chapter 6
The economic case for agroforestry

Market opportunities – direct outputs from the tree component of agroforestry systems

By Clive Thomas, Senior Policy Adviser
(Forestry & International Land Use), Soil Association

As has been established in earlier chapters, the economic case for agroforestry can be considered in three main ways:

1. **The value of enhanced ecosystem services from agroforestry systems**
2. **The value of enhanced agriculture outputs from agroforestry systems**
3. **The value that can be derived directly from the tree component of agroforestry systems.**

This section considers in more detail some of the opportunities and considerations in respect of this final set of values, which for simplicity we will refer to as the 'tree outputs'. This information should be viewed as general guidance only, based on well-established and understood forestry principles that are likely to have some applicability. However, so far there are only a few UK agroforestry examples that can be studied in detail, to test to what degree the deep understanding of tree performance in UK forestry conditions might apply to a range of agroforestry systems and situations.

Figure 45: **Timber waiting to be processed and woodchip**

©Soil Association

Chapter 6
The economic case for agroforestry

Market considerations

Many of the 'tree outputs' have long-established markets such as timber, fuel and food. Others such as recreation and leisure or carbon storage and more novel non-timber forest products, like foliage, biochar and Christmas trees, offer a developing or niche opportunity for farm enterprises

Timber markets

There is a significant forest industry in the UK, with established markets for a range of timber species. Sawmills and other processors can cater for small-diameter softwood material from conifers for fencing material or pulp/chipwood production, to larger-diameter sawlog material for construction and carcassing material. More specialist sawmills cater for high-grade hardwood logs for planking or veneer and poplar for packaging and carcassing. For some timber, there may only be a few options, e.g. cricket bat willow, furniture-grade material or coppice such as sweet chestnut paling or hazel hurdles, but these niche markets often attract a premium for good quality material.

Food (from trees) market

Markets are well established for apples, pears, plums, cherries etc. Whereas considerable potential exists for hazelnuts/cobnuts (squirrel control required), walnuts and chestnuts (the warming climate will suit these), but they have long establishment periods and, in some situations, walnuts can inhibit the growth of neighbouring plants.

Woodfuel market

Wood for fuel has seen a significant resurgence in recent years, stimulated by both the renewable energy market and the aesthetic aspects of wood burning. Depending on the species grown, well-seasoned firewood, split to size and delivered, can be a high-end product demanding a high market price. Even low-grade, unseasoned material can find a market, although location is usually a critical factor as transport costs can be high.

Other non-timber forest products

Niche markets exist for foliage for floristry and the Christmas market, e.g. holly and mistletoe, as well as Christmas trees as a stand-alone option. There are examples of silvopastoral systems growing Christmas trees successfully with sheep, if the chosen tree species is not too palatable, the stocking density of the sheep is not too high, and the sheep are excluded whilst the trees are actively growing.

Carbon finance

The carbon market in the UK currently only exists as a voluntary market, however the Forestry Commission and other stakeholders have developed a Woodland Carbon Code to establish a common set of rules and standards for any investment that is registered. Investors have been prepared to buy the verified carbon credits that can be claimed under the Code and this market is expected to develop further in the UK based on government support and requirements for UK companies to report their greenhouse gas emissions.

Currently the Woodland Carbon Code assumes that the land is classed as woodland so only agroforestry established by reducing stand density in woodland to create agroforestry would apply. For agroforestry operating on land classified as agricultural it is not currently possible to carbon trade under CAP rules, though this may change in the future. More information is available here https://woodlandcarboncode.org.uk/.

Recreation and leisure market

There is an increasing number of examples of the tree component on a farm enterprise providing a recreation or leisure market opportunity. Examples include camping, glamping, paintballing, bushcrafts/woodcrafts and forest schools.

Substitution opportunities

As well as direct sales of 'tree outputs', the opportunity to substitute tree outputs grown on the farm enterprise for products that are usually bought in can offer significant economic benefit. Examples may include simple opportunities such as the use of timber outputs for on-farm fencing requirements or when the trees are more mature, opportunities for farm building renovation, repairs or new build using on-farm harvested timber beams or cladding, for instance. Other substitution opportunities may require some initial investment, to support a longer-term substitution opportunity, e.g. replacement of an oil boiler with a woodfuel boiler, which is then fired by on-farm woodfuel production.

Tree establishment, maintenance and management

Good establishment will lead to healthy, robust and productive trees. This may require excluding stock completely during the establishment phase. Beyond this basic tenet, there are few special considerations over and above those already described in earlier chapters for ensuring that the tree outputs described above are realised. Whether the trees are grown primarily for agriculture or ecosystems service outputs or specifically for tree outputs, the basic rules of procuring good-quality stock of known origin, good planting practice and effective tree protection

Chapter 6
The economic case for agroforestry

are essential. The following listed provides a few specifics and general reminders:

Planting stock

It is always important to purchase good-quality planting stock, from a registered tree nursery that can supply evidence of origin. However, if tree outputs such as timber and food (fruit, nuts and berries) are a major consideration, then even more due diligence should be paid to the provenance of the stock, which will help to ensure that the young trees will go on to perform as expected.

Establishment

In general, the smaller the planting stock the quicker and easier it is to establish, e.g. whips and transplants establish much more quickly than standard trees and will often quickly overtake these bigger initial plants after a few growing seasons. There is the added benefit that smaller planting stock will always be cheaper than larger trees, so standards should only be considered in exceptional circumstances. However, whether smaller or larger stock is used, all trees should be handled with care and kept heeled into a temporary trench or stored in planting bags in cool conditions prior to planting. Ideally plant your trees in the autumn or early spring. Never plant in the summer or when the ground is very cold, or frost is likely. Initial vegetation control for approximately 100 cms diameter around each tree, ideally via a mulch, will be essential for effective establishment. If all these measures are in place and the trees are planted correctly, unless drought occurs, there should be no need to provide artificial watering, even in the first growing season.

Tree protection

Livestock and other natural browsers (deer, rabbits, hares) are the single biggest threat to young trees and if these factors can be controlled for the first five to 10 years, then the forestry component of the agroforestry systems will establish quickly and then be far more resilient and robust.

As has already been stated, planting small initial planting stock is preferred, so full stock exclusion using some form of fencing for this phase will be essential. Some trees with palatable bark may always require protection, and at all ages, in a silvopastoral system. This is also likely to take the form of some type of permanent fencing system, either for lines or groups of trees or individual fencing guards. Other tree species which can resist some browsing pressure or have unpalatable bark will usually only require protection until they are established and the main growing tip of the tree is above browsing height and the tree is strong enough to not be damaged by sheep/cattle/horses rubbing and pushing the tree. Tree species that require long-term individual tree protection are unlikely to be well suited to silvopastoral systems and alternative species should be chosen.

Tree management

After the establishment phase, trees in general are self-sufficient and require little management intervention. However, if the tree outputs include timber and food (fruit, nuts and berries), then pruning is likely to be required to maximise the quality and form of the timber that is grown or, depending on the tree species, maximise the harvest of fruit, nuts and berries. Tree pruning for timber quality includes singling out multiple leaders, so that the tree grows straight (formative pruning) and removing side branches to reduce knot size and/or either improve the strength properties of the timber or improve the aesthetic value through a cleaner appearance (high pruning). Both can add significant value to any harvested timber that is sold into a market.

Harvesting

There is a very wide range of harvesting activity that might be required, ranging from small-scale removal of thinnings from densely planted trees or the cutting of coppice, through to the felling of mature trees upwards of five tonnes in weight. Health and safety of those carrying out the operations and any other occupants of the land (including livestock) is of paramount importance across the full spectrum of harvesting activity but especially relevant when larger trees are felled, which will require qualified specialists, training, skills and equipment.

On-farm processing

The tree outputs that are harvested may either be sold direct to a market or some form of on-farm processing may occur, primarily to add value for either a market or substitution opportunity. Mobile sawbenches can process timber for on-farm use, such as fencing, cladding or construction lengths. Fuel and firewood production often lends itself well to on-farm drying, splitting or chipping and means that if sold a higher-value product is transported rather than low-grade material in bulk.

Figure 46: **De-limbing a felled timber tree**

©Soil Association

Tree growth rate yield and products

A challenge for agroforestry is that as a hybrid system combining woody forestry perennials and agricultural crops and land use, different terminology, language and yield measurement approaches are used in agriculture and forestry. Depending on the dominant output, an agroforester may wish to use an agricultural measure of output, i.e t/ha, or a forestry measure of output, i.e timber yield class (YC).

The growth of a tree may be measured in terms of height, diameter, volume or weight, but volume is the most meaningful for management purposes, and is expressed in forestry as the yield class system.

In an even-aged stand of trees, the cumulative volume production divided by the age of the stand is referred to as the mean annual increment (MAI). However, the growth curve of a tree is not a straight line. During the early years, growth is vigorous, but growth reaches a maximum and then declines with age. The point at which the MAI curve reaches its maximum is the maximum average rate of volume increment that the stand can achieve, and this number is the yield class. Therefore, a stand with a maximum MAI of 20 m^3 per ha has a yield class of 20.

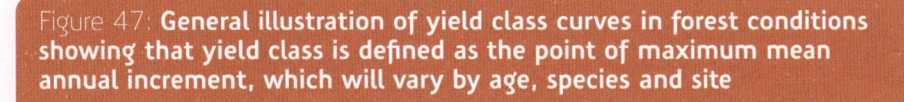

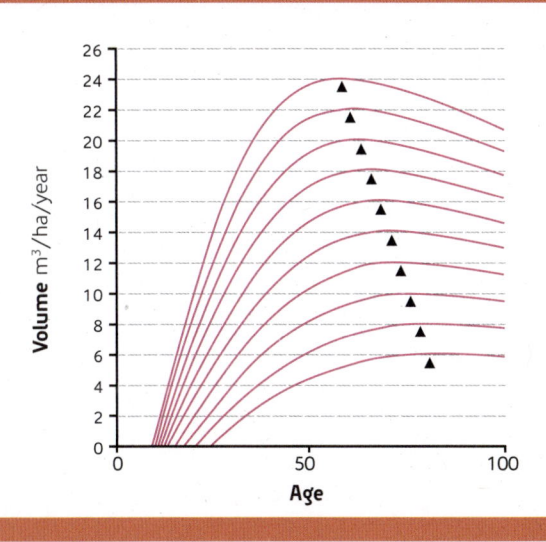

Figure 47: **General illustration of yield class curves in forest conditions showing that yield class is defined as the point of maximum mean annual increment, which will vary by age, species and site**

Chapter 6
The economic case for agroforestry

It should be remembered that the yield class system has been developed based on measurements in even aged-stands of trees planted at relatively close spacing in a forest situation, meaning that canopy closure occurs at a relatively young age. The initial and final tree spacing in agroforestry systems are likely to be far wider than even the widest 3.0 m spacing in a forest context (1,100 stems per ha), and therefore yield class can only be a guide at best for productivity of trees in agroforestry systems.

However, the general ground rules that yield class will vary between species on any given site and between sites based on fertility, water availability, elevation and exposure etc. will apply and therefore give a basis for estimating the productivity of different tree species on different sites. Most broadleaf species (except poplars) will typically achieve yield classes of 2–12 at best in optimal forest conditions, with poplar generally achieving YC 10–12 and most conifers performing in the range of YC 8 up to 24 for the very best performing species, on the most productive sites in forest conditions.

In a forestry situation, most volume is accumulated on the tree stem as branches are kept small due to the shade of close spacing. At wider spacings in agroforestry systems, there will be more accumulation of volume in branchwood rather than stemwood.

In any situation, a tree stem can be visualised as a tapering cylinder with a new layer accumulated each growing season. When this tapering cylinder is felled and trimmed, the options for what can be cut from the length of the tree are greatly affected by the diameter of the cylinder, as well as the taper, straightness and overall length of the cylinder. Once the cylinder (stem) diameter gets to a certain size, the options for cutting certain sizes and grades of timber increases greatly, as a larger cross-section of solid timber is available. Therefore, when trees are small, and the stem has a small diameter, fencing posts and woodfuel may be the only options. When the trees are older, and the stem diameter is bigger, standard sizes of timber become an option and when the diameter reaches a certain size, wide planks can be cut.

However, volume will only be one consideration and depending on the objectives for growing trees, in both a forest and agroforestry context, the quality of the timber will sometimes be more important than yield. Although there are exceptions, as a general principle there is some correlation between durability and strength, as two aspects of timber quality and growth rate, with faster growing species tending to produce less durable timber with lower strength properties.

Tree species selection

Different species of tree offer different options for tree outputs from agroforestry systems. Listed in Table 18 are some of the main options, with a brief indicator of some of the key opportunities and considerations based on forestry practice, again due to the current lack of UK agroforestry examples. In any case, specialist advice should always be sought before making a species selection, as a good understanding of the site and its potential for different options will be fundamental to success. In addition, many of the tree species listed remain largely untested in a UK agroforestry context, so local expertise and knowledge will also be important in making a judgement of what to plant and where to plant it.

With the UK's climate changing, the suitable range of some tree species currently only grown in South East England will extend north and west. It may therefore be worth considering using tree species grown in other parts of the temperate world. Southern England especially Kent may end up with an agro-climate similar to Bordeaux in the next 50 years for instance. For a full treatment of trees in temperate agroforestry across the world it is worth consulting Gordon et al 2018[9].

Compared to some other European countries, the UK only has a small woodland economy at the farm-scale, so it is worth considering finding a partner with timber experience if you decide to grow for this market.

The best option for farmers with small-scale production might be to work in partnership with forestry-scale operators and/or in co-operatives. Actual market opportunities will vary and therefore it will be best to take advice on potential sale price, percentage of saleable timber that might be extracted from the trees in your agro-forestry system, the timber quality and the minimum number of trees to make extraction worthwhile.

The John Nix Farm Management Pocketbook[1] gives general guidance on prices including hardwoods and the following website shows the price size curves for different hardwoods https://www.growninbritain.org/selling-hardwood-trees-2/

Finally, some of the markets, such as woodfuel, recreation and leisure and carbon are less sensitive to specific tree species and will usually work best as a secondary output alongside timber and/or food.

Chapter 6
The economic case for agroforestry

Table 18: Considerations for choosing which tree to plant

Tree species	Outputs (primary)	Pastoral or arable[1]	Upland or lowland[2]	Special considerations
Oak	Timber, woodfuel	Pastoral or arable	Lowland or upland	Opportunity for high-grade timber but with very long rotation, although opportunities for woodfuel from year 20. Careful attention when choosing provenance and high pruning will be required.
Poplar	Timber	Pastoral or arable	Lowland	Opportunity for quality timber production on short-rotations. Easy to establish. Pruning required. Established markets, although few on-farm opportunities for substitution.
Douglas fir and other redwoods e.g. Western red cedar	Timber – construction, fencing	Pastoral	Lowland or upland	Opportunity for quality timber production on average rotations. Requires careful site selection (free-draining soils and avoid frost hollows). Established markets and good opportunities for substitution (naturally durable so no need for treatment), e.g. fencing material, construction, cladding.
Hazel	Fencing, woodfuel, fruit	Pastoral or arable	Lowland	Has potential for a range of useful products grown via a coppice system, especially fencing (hurdles). Cobnut production is likely to benefit from a warming climate. Susceptible to squirrel damage.
Pines	Timber, fencing	Pastoral or arable	Lowland or upland	Opportunity for timber production on average rotations. Tolerant of drier soils in lowland and upland situations. Established markets and some opportunities for substitution, although will need treatment to ensure durability in outdoor situations, e.g. fencing material, construction, cladding.
Sycamore	Timber, woodfuel	Pastoral or arable	Lowland or upland	Along with oak has highest value as a sawlog, but requires seasonal felling, rapid transport to market and is also very susceptible to squirrel damage. Opportunities for woodfuel from year 20. Will tolerate upland conditions and moderate exposure, although unlikely to yield quality timber in these situations.

Beech	Timber, woodfuel	Pastoral	Lowland or upland	Has potential for high-value timber in some situations but very susceptible to squirrel damage. Opportunities for woodfuel from year 20 and due to dense canopy, probably not suitable for arable systems but great potential for semi-upland shelter as will grow at reasonable elevation.
Cricket bat willow	Timber	Pastoral	Lowland	Grown as 'wood pasture' on relatively short rotations of 12–20 years. With strict pruning requirements. (Discuss with specialist buyers before establishment.)
Walnut (European & black)	Timber, fruit	Pastoral or arable	Lowland	Opportunity for high-grade timber but with a long rotation. Careful attention when choosing site and provenance for timber and nut production. High pruning will be required for timber and nut yield and quality will potentially benefit from a warming climate.
Wild cherry	Timber, fruit	Pastoral or arable	Lowland	Opportunity for high-grade timber on shorter rotations. Careful attention when choosing site, provenance and high pruning will be required. Wild cherry is unlikely to be suitable for fruit production and is usually best grown alongside other tree species.
Sweet chestnut	Timber, fencing, woodfuel, fruit	Pastoral or arable	Lowland	Has potential for a range of useful products, especially fencing and nut production and is likely to benefit from a warming climate. Susceptible to squirrel damage. Opportunities for woodfuel from year 20 and due to dense canopy, wide spacings will be required in arable situations.
Apple, pear, cherry, plum etc.	Fruit, woodfuel	Pastoral or arable	Lowland	Primary objective will be fruit production and this should determine site choice and variety. Some opportunities for woodfuel from pruning and when fruit production declines.

[1] It should be noted that many of these species remain largely untested on any scale in a UK agroforestry context. Therefore, these are suggestions as to which agroforestry system may have most potential for these species

[2] In general, the higher upland situations will be unsuitable for tree growth and therefore agroforestry systems. The upland sites with most potential will be those on slopes, and/or with free-draining soils, not too exposed. Micro-site considerations will be very important when selecting system and species

References

CHAPTER 1

1. Forestry Commission (2017). Tree cover outside woodland in Great Britain National Forest Inventory. April 2017. https://www.forestresearch.gov.uk/documents/2699/FR_Tree_cover_outside_woodland_in_GB_statistical_report_2017.pdf
2. Upson MA, Burgess, PJ, Morison, JIL (2016). Soil carbon changes after establishing woodland and agroforestry trees in a grazed pasture. Geoderma 283: 10-20.
3. Burgess PJ, Rosati A (2018). Advances in European agroforestry: Results from the AGFORWARD project. Agroforestry Systems 92:801-810. https://doi.org/10.1007/s10457-018-0261-3
4. Mosquera-Losada MR, Santiago Freijanes JJ, Pisanelli A, Rois M, Smith J, den Herder M, Moreno G, Malignier N, Mirazo JR, Lamersdorf N, Ferreiro-Domínguez N, Balaguer F, Pantera A, Rigueiro-Rodríguez A, Gonzalez-Hernández P, Fernández-Lorenzo JL, Romero-Franco R, Chalmin A, Garcia de Jalon S, Garnett K, Graves A, Burgess PJ (2016). Extent and success of current policy measures to promote agroforestry across Europe. Deliverable 8.23 for EU FP7 Research Project: AGFORWARD 613520. (8 December 2016). 95 pp. http://www.agforward.eu/index.php/en/extent-and-success-of-current-policy-measures-to-promote-agroforestry-across-europe.html
5. Lawson GJ, Brunori A, Palma JHN, Balaguer P (2016) Sustainable management criteria for agroforestry in the European Union. In 3rd European Agroforestry Conference 2016 Book of Abstracts 375-378. (Eds. Gosme M et al.). 23-25 May 2016, Montpellier SupAgro, France.
6. Rodwell JS, Paterson G (1994). Creating New Native Woodlands. Forestry Commission Bulletin 112, HMSO, London
7. Forestry Commission Scotland. Woodland grazing toolbox. https://forestry.gov.scot/woodland-grazing-toolbox
8. Plieninger T, Hartel T, Martín-López B, Beaufoy G, Bergmeier E, Kirby K, Montero MJ, Moreno G, Oteros-Rozas E, van Uytvanck J (2015). Wood-pastures of Europe: Geographic coverage, social-ecological values, conservation management, and policy implications. Biological Conservation 190, 70-79.
9. Read H (2000). What are veteran trees and why are they important? Veteran Trees: A guide to good management. Natural England, pp. 13-24.
10. den Herder M, Moreno G, Mosquera-Losada RM, Palma JHN, Sidiropoulou A, Santiago Freijanes JJ, Crous-Duran J, Paulo JA, Tomé M, Pantera A, Papanastasis VP, Mantzanas K, Pachana P, Papadopoulos A, Plieninger T, Burgess PJ (2017) Current extent and stratification of agroforestry in the European Union. Agriculture, Ecosystems and Environment 241: 121-132.
11. Agroforestry Research Trust (2018). Forest farming. https://www.agroforestry.co.uk/about-agroforestry/forest-farming/
12. Ministerial Conference on the Protection of forests in Europe (2015). Summary for Policy Makers: State of Europe's forests 2015. https://foresteurope.org/wp-content/uploads/2016/08/summary-policy-makers.pdf
13. Torralba, M., Fagerholm, N., Burgess, P.J., Moreno, G., Plieninger, T. (2016). Do European agroforestry systems enhance biodiversity and ecosystem services? A meta-analysis. Agriculture, Ecosystems and Environment 230: 150-161.

CHAPTER 2

1. Marshall, et al., (2014). The impact of rural land management changes on soil hydraulic properties and runoff processes: Results from experimental plots. Hydrological Processes, 28, 2617-2629.
2. Newman SM (1986). A Pear and Vegetable Interculture System: Land Equivalent Ratio, Light Use Efficiency and Dry Matter Productivity. Experimental Agriculture 22:383-392.
3. Redman, G. (ed) (2018). John Nix Pocketbook for Farm Management 2019, 49th Edition, Melton Mowbray, Agro Business Consultants.
4. Ecological Site Classification Decision Support System (ESC-DSS), Forestry Commission https://www.forestresearch.gov.uk/tools-and-resources/forest-planning-and-management-services/ecological-site-classification-decision-support-system-esc-dss/
5. Hoare AH (1928) The English Grass Orchard and the Principles of Fruit Growing. Ernest Benn, London.
6. Crawford, M. (2010) Creating a Forest Garden. Green Books ISBN 978 1 900322 62
7. Doyle, C. J., Evans, J. and Rossiter J (1986) Agroforestry: An economic appraisal of the benefits of intercropping trees with grassland in lowland Britain. Agricultural Systems 21: 1-32.
8. Burgess, P.J., Incoll, L.D., Corry, D.T., Beaton, A. & Hart, B.J. (2005). Poplar growth and crop yields within a silvoarable agroforestry system at three lowland sites in England. Agroforestry Systems 63(2): 157-169. http://hdl.handle.net/1826/872
9. Burgess, P.J., Incoll, L.D., Hart, B.J., Beaton, A., Piper, R.W., Seymour, I., Reynolds, F.H., Wright, C., Pilbeam & Graves, A.R.. (2003). The Impact of Silvoarable Agroforestry with Poplar on Farm Profitability and Biological Diversity. Final Report to DEFRA. Project Code: AFO105. Silsoe, Bedfordshire: Cranfield University. 63 pp. http://sciencesearch.defra.gov.uk/Document.aspx?Document=AFO105_7417_FRP.pdf
10. Newman, S.M., Wainwright, J., Oliver, P.N. and Acworth, J.M. (1991b) Walnut agroforestry in the UK: Research 1900-1991 assessed in relation to experience in other countries. Proceedings of the 2nd Conference on Agroforestry in North America, Springfield, Missouri, pp. 74-94.
11. Newman, S.M., Park, J., Wainwright, J., Oliver, P., Acworth, J.M. and Hutton, N. (1991c) Tree productivity, economics and light use efficiency of poplar silvoarable systems for energy. Proceedings of the 6th European Conference on Biomass Energy Industry and Environment, Athens.
12. Newman, S.M. (2018) Rural new settlements in 2018 – What might Ebenezer Howard say? Journal of the Town and Country Planning Association August 2018. pp 301-306
13. Newman, S.M. (1985). An Investigation of the Feasibility of Combined Energy Cropping and Landscape Management. In Egneus, H. and Ellegard, A. (Eds) Bioenergy 84: Proceedings of an International Conference on Energy from Biomass Gotteborg 1984 Volume 11 Biomass Resources: 159-161. Elsevier Applied Science London
14. Tree I. (2018) Wilding: The return of nature to a British farm. Picador Publishing.

CHAPTER 3

1. den Herder M, et al. (2016). Current extent and trends of agroforestry in the EU27. Deliverable Report 1.2 for EU FP7 Research Project: AGFORWARD 613520. (15 August 2016). 2nd Edition. 76 pp.
2. Millett J, Godbold D, Smith AR, Grant H (2011) Nitrogen fixation and cycling in Alnus glutinosa, Betula pendula and Fagus sylvatica woodland exposed to free air CO_2 enrichment. Oecologia (2012) 169:541-552
3. Blore D (1994) Benefits of remnant vegetation: Focus on rural lands and rural communities. Paper prepared for the 'Protecting remnant bushland' seminar, Orange Agricultural College, October

1994.
4. Broster, et al. (2012). Evaluating seasonal risk and the potential for windspeed reductions to reduce chill index at six locations using GrassGro. Animal production science.52(10):921-8
5. Key N, Sneeringer S, and Marquardt D (2014). Climate Change, Heat Stress, and U.S. Dairy Production, ERR-175, U.S. Department of Agriculture, Economic Research Service, September 2014.
6. Blackshaw JK, and Blackshaw AW (1994) Heat stress in cattle and effect of shade on production and behaviour: a review. Aust. J. Exp. Agr. 34:285-295.
7. The Woodland Trust. The role of trees in free range poultry farming. https://www.woodlandtrust.org.uk/mediafile/100256995/The-role-of-trees-in-free-range-poultry-farming.pdf
8. Luske B et al. 'Agroforestry for ruminants in the Netherlands', AGFORWARD 5.14, 15 August 2017. Available online: https://www.agforward.eu/index.php/en/fodder-trees-for-cattle-and-goats-in-the-netherlands.html
9. Burgess PJ, Chinery F, Eriksson G, Pershagen E, Pérez-Casenave C, Lopez Bernal A, Upson A, Garcia de Jalon S, Giannitsopoulos M, Graves A (2017). Lessons learnt – Grazed orchards in England and Wales. AGFORWARD project. 21 pp.
10. Forestry Commission Scotland. Sheep and Trees. https://forestry.gov.scot/support-regulations/sheep-and-trees
11. Améndola L, Solorio F J, González-Rebeles C, Galindo F (2013) Behavioural indicators of cattle welfare in silvopastoral systems in the tropics of México. In: (Eds.) M J Hötzel, L C P M Filho. Proceedings of 47th Congress of International Society for Applied Ethology, 2-6 June 2013, Florianópolis, Brazil. p.150.
12. Forestry Commission Scotland. Woodland grazing toolbox. https://forestry.gov.scot/woodland-grazing-toolbox
13. Forestry Commission Scotland. Woodland grazing and pigs. https://forestry.gov.scot/woodland-grazing-toolbox/grazing-management/grazing-regime/selecting-species-and-breed/pigs
14. Alberta Agriculture and Rural development https://www1.agric.gov.ab.ca/$Department/deptdocs.nsf/all/agdex12072
15. The Woodland Trust. The role of trees in sheep farming - Integrating trees to boost production and improve animal health, https://www.woodlandtrust.org.uk/publications/2018/07/the-role-of-trees-in-sheep-farming/

CHAPTER 4

1. García de Jalón S, Burgess PJ, Graves A, Moreno G, McAdam J, Pottier E, Novak S, Bondesan V, Mosquera-Losada MR, Crous-Durán J, Palma JHN, Paulo JA, Oliveira TS, Cirou E, Hannachi Y, Pantera A, Wartelle R, Kay S, Malignier N, Van Lerberghe P, Tsonkova P, Mirck J, Rois M, Kongsted AG, Thenail C, Luske B, Berg S, Gosme M, Vityi A (2018). How is agroforestry perceived in Europe? An assessment of positive and negative aspects among stakeholders. Agroforestry Systems 92:829–848. https://doi.org/10.1007/s10457-017-0116-3
2. Graves AR, Morris J, Deeks LK, Rickson RJ, Kibblewhite MG, Harris JA, Farewell TS, Truckell I (2015). The total costs of soil degradation in England and Wales. Ecol. Econ, 119: 399–413.
3. Wright C (1994). The distribution and abundance of small mammals in a silvoarable agroforestry system. Agroforestry Forum 5 (2): 26–28
4. Wartelle R, Mézière D, Gosme M, Ia-Laurent L (2016) System report: Weed Survey in Northern Silvoarable Group in France 15 January 2016 http://www.agforward.eu/index.php/en/agroforestry-for-arable-farmers-in-northern-france.html
5. Kanzler, M., Böhm, C., Mirck, J., Schmitt, D., Veste M. (2018). Microclimate effects on evaporation and winter wheat (Triticum aestivum L.) yield within a temperate agroforestry system. Agroforestry Systems. https://doi.

org/10.1007/s10457-018-0289-4
6. Van Lerberghe P (2017). Agroforestry Best Practice leaflet 4: Planning an agroforestry project. AGFORWARD project. 2 pp.
7. Burgess PJ, Incoll LD, Hart BJ, Beaton A, Piper RW, Seymour I, Reynolds FH, Wright C, Pilbeam D, Graves AR (2003). The Impact of Silvoarable Agroforestry with Poplar on Farm Profitability and Biological Diversity. Final Report to DEFRA. Project Code: AF0105. Silsoe, Bedfordshire: Cranfield University. 63 pp.
8. Paris P, Dalla Valle C (2017). Agroforestry Innovation leaflet 32: Hybrid poplar and oak along drainage ditches. AGFORWARD project. 2 pp. http://www.agforward.eu/index.php/en/trees-for-timber-intercropped-with-cereals-445.html
9. Dupraz C, Blitz-Frayret C, Lecomte L, Molto Q, Reyes F, Gosme M (2018). Influence of latitude on the light availability for intercrops in an agroforestry alley-cropping system. Agroforestry Systems 92: 1019-1033.
10. Incoll L, Newman S (2000). Arable crops in agroforestry systems. Agroforestry in the UK. 71-80. (Eds Hislop M, Claridge J). Forestry Commission Bulletin 122.
11. Pasturel P (2004). Light and water use in a poplar silvoarable system. Unpublished MSc by Research Thesis, Cranfield University. 143 pp.
12. Wartelle R, Mézière D, Gosme M, Ia-Laurent L, Muller L (2017). Lessons learnt: Weeds and silvoarable agroforestry in Northern France. 13 November 2017. 9 pp. Available online: http://agforward.eu/index.php/en/agroforestry-for-arable-farmers-in-northern-france.html
13. Smith J, Westaway S, Venot C, Cathcart-James M (2017b). Lessons learnt: Silvoarable agroforestry in the UK (Part 2). 8 September 2017. 18 pp. Available online: http://www.agforward.eu/index.php/en/silvoarable-agroforestry-in-the-uk.html
14. Griffith, J., Phillips, D.S., Compton, S.G., Wright, C., Incoll, L.D. (1998). Slug number and slug damage in a silvoarable agroforestry landscape. Journal of Applied Ecology 35, 252-260.
15. Graves AR, Burgess PJ, Palma JHN, Herzog F, Moreno G, Bertomeu M, Dupraz C, Liagre F, Keesman K, van der Werf W, Koeffeman de Nooy A, van den Briel JP (2007). Development and application of bio-economic modelling to compare silvoarable, arable and forestry systems in three European countries. Ecological Engineering 29: 434-449.
16. Burgess PJ, Incoll LD, Corry DT, Beaton A, Hart BJ (2005). Poplar growth and crop yields within a silvoarable agroforestry system at three lowland sites in England. Agroforestry Systems 63: 157-169.

CHAPTER 5

1. Hislop M, Gardiner B, Palmer H (2006). The principles of wood for shelter. Forestry Commission Information Note
2. HedgeLink, The Hedgerow Management Cycle. Available from: http://www.hedgelink.org.uk/cms/cms_content/files/78_hedgelink_a5_T2pp_leaflet_7.pdf
3. Lofti A, Javelle A, Baudry J, Burel F. (2010). Interdisciplinary Analysis of Hedgerow Network Landscapes' Sustainability. Landscape Research 35:415Ð426.
4. Wolton RJ (2012). The yield and cost of harvesting wood fuel from hedges in South-West England. Unpublished report to the Tamar Valley and Blackdown Hills AONBs. Locks Park Farm, Hatherleigh, Okehampton, Devon, EX20 3LZ
5. Chambers, M., Crossland, M., Westaway, S., Smith, J. (2015) A guide to harvesting woodfuel from hedges. ORC Technical Guide. http://tinyurl/TWECOM
6. Agriculture and Horticulture Development Board (2018) The Bedding Materials Directory http://beefandlamb.ahdb.org.uk/wp-content/uploads/2018/10/Bedding-materials-directory.pdf

7. Woodland Trust (2016) Keeping Rivers Cool: A Guidance Manual – Creating riparian shade for climate change adaptation. http://www.woodlandtrust.org.uk/mediafile/100814410/pg-wt-060216-keeping-rivers-cool.pdf

CHAPTER 6

1. Redman, G. (ed) (2018). John Nix Pocketbook for Farm Management 2019, 49th Edition, Melton Mowbray, Agro Business Consultants.
2. Organic Farm Management Handbook 11th Edition (January 2017), N Lampkin, M Measures, S Padel
3. Macaulay Land Use Research Institute, http://macaulay.webarchive.hutton.ac.uk/fmd/agroforest.html
4. Graves, A.R., Burgess, P.J., Palma, J.H.N., Herzog, F., Moreno, G., Bertomeu, M., Dupraz, C., Liagre, F., Keesman, K., van der Werf, W. Koeffeman de Nooy, A. & van den Briel, J.P. (2007). Development and application of bio-economic modelling to compare silvoarable, arable and forestry systems in three European countries. Ecological Engineering 29: 434-449.
5. SAFE: Silvoarable Agroforestry For Europe https://www1.montpellier.inra.fr/safe/english/index-report.htm
6. http://www.eurafagroforestry.eu/afinet/rains/agroforestry-action/whitehall_farm_an_innovative_silvoarable_orchard_system_in_the_UK
7. https://www.dartington.org/about/our-land/agroforestry/
8. The Woodland Trust. Tree planting and farming hand in hand: How to plant without affecting your subsidies. https://www.woodlandtrust.org.uk/publications/2016/11/allerton-project/
9. Gordon AM, Newman SM, and Coleman BRW (2018). Temperate Agroforestry Systems 2nd Edition CAB International Boston MA. ISBN 9781780644851

Authors

Dr Paul Burgess
Paul is Reader in Crop Ecology and Management at Cranfield University in Bedfordshire in England. He was Co-ordinator of a European agroforestry project called AGFORWARD that ran from 2014 to 2017 and is the current Secretary of the Farm Woodland Forum.

Prof. Steven Newman
Steven is MD of BioDiversity International Ltd. The company is involved with investments in agroforestry worldwide leading to the planting of over eight million trees. He is currently Visiting Professor in the School of Biology at Leeds University. He is still involved in research, consultancy and partnerships linked to UK agroforestry trials he set up with farmers.

Dr Tim Pagella
Tim works at Bangor University in North Wales. He coordinates the teaching of agroforestry at the university and is involved in agroforestry research both in the UK (mainly with silvopastoral systems) and in many parts of Africa and Asia (working with the World Agroforestry Centre - ICRAF).

Dr Jo Smith
Jo is a Principal Researcher leading the Agroforestry programme at the Organic Research Centre. With a background in soil biodiversity and agri-environment schemes, Jo works on European projects investigating agroforestry as a way of reconciling production with protection of the environment.

Sally Westaway
Sally is a Senior Agroforestry Researcher at the Organic Research Centre. Her on-farm research focusses on the role of trees and hedges in the farmed environment and methods to enhance the agronomic, environmental and economic performance of agroforestry systems.

Stephen Briggs MSc, BSc, NSch
Stephen is a practical farmer, consultant and Soil and Water Manager at Innovation for Agriculture. As a Nuffield Scholar in 2011 Stephen studied agroforestry around the world and pioneered largescale silvoarable apple production on his own organic farm in Cambridgeshire.

Ian Knight
Ian is an organic farmer, agronomist and farm consultant working as a Director at Abacus Agriculture. Since 2012 Ian has been responsible for the successful delivery of agroforestry research and training projects in the UK and across Europe.

Dr Lindsay Whistance
Lindsay is a Senior Livestock Researcher at the Organic Research Centre, researching animal behaviour, health and welfare. Her activities include investigating the role of diverse, species-rich environments, including agroforestry, as a habitat and a source of food, medicine and pain relief for domestic animals.

Editors

Ben Raskin
Ben has worked in horticulture for 25 years and has a wide range of practical commercial growing experience. He is Head of Horticulture for the Soil Association and leads on their Agroforestry work. He is also currently implementing a 200-acre agroforestry planting in Wiltshire.

Simone Osborn
Simone is a project manager in the Producer Support team at the Soil Association supporting a variety of projects. She has a background in project management, account management and publishing.

Acknowledgements

We would like to gratefully acknowledge the support of all who have made this edition of the Agroforestry Handbook possible. Thank you to the John Ellerman Foundation who have provided the funding to produce this publication. Thank you to all the farmers and researchers whose work and experimentation underpins the current body of knowledge, in particular:

- The Woodland Trust, David Brass – The Lakes Free Range Egg Company, Stephen and Lynn Briggs – Whitehall Farm, and Jonathan Francis – Tyn-Yr-Wtra Farm for use of case studies
- Clive Thomas (FICFor), Senior Policy Adviser (Forestry & International Land Use), Soil Association, for pp. 132–141 in Chapter 6
- Also for Chapter 6 we would like to gratefully acknowledge the help of Jez Ralph, Timber Strategies for his invaluable information and insight into the commercial forestry sector relative to agroforestry. In addition, our special thanks go to Graham Redman from The Anderson Centre for allowing us to utilise the John Nix Farm Management Pocketbook, and to the Organic Research Centre for allowing us to use data from the Organic Farm Management Handbook.

Design by Andrew Evans Graphic Design – evansgraphic.co.uk

Printing by Wells Printing Limited, Unit 15 Bath Business Park, Foxcote Avenue, Peasedown St John, Bath BA2 8SF